D TECHNOLOGIES

ND

website

website

ABLE POLYMERS
INABLE POLYMERS
OPOL-2009)

MATERIALS SCIENCE AND TECHNOLOGIES

Additional books in this series can be found on Nova's
under the Series tab.

Additional E-books in this series can be found on Nova's
under the E-books tab.

MATERIALS SCIENCE AND TECHNOLOGIES

BIODEGRADABLE POLYMERS AND SUSTAINABLE POLYMERS (BIOPOL-2009)

ALFONSO JIMENEZ
AND
G.E. ZAIKOV
EDITORS

Nova Science Publishers, Inc.
New York

Copyright © 2011 by Nova Science Publishers, Inc.

All rights reserved. No part of this book may be reproduced, stored in a retrieval system or transmitted in any form or by any means: electronic, electrostatic, magnetic, tape, mechanical photocopying, recording or otherwise without the written permission of the Publisher.

For permission to use material from this book please contact us:
Telephone 631-231-7269; Fax 631-231-8175
Web Site: http://www.novapublishers.com

NOTICE TO THE READER

The Publisher has taken reasonable care in the preparation of this book, but makes no expressed or implied warranty of any kind and assumes no responsibility for any errors or omissions. No liability is assumed for incidental or consequential damages in connection with or arising out of information contained in this book. The Publisher shall not be liable for any special, consequential, or exemplary damages resulting, in whole or in part, from the readers' use of, or reliance upon, this material. Any parts of this book based on government reports are so indicated and copyright is claimed for those parts to the extent applicable to compilations of such works.

Independent verification should be sought for any data, advice or recommendations contained in this book. In addition, no responsibility is assumed by the publisher for any injury and/or damage to persons or property arising from any methods, products, instructions, ideas or otherwise contained in this publication.

This publication is designed to provide accurate and authoritative information with regard to the subject matter covered herein. It is sold with the clear understanding that the Publisher is not engaged in rendering legal or any other professional services. If legal or any other expert assistance is required, the services of a competent person should be sought. FROM A DECLARATION OF PARTICIPANTS JOINTLY ADOPTED BY A COMMITTEE OF THE AMERICAN BAR ASSOCIATION AND A COMMITTEE OF PUBLISHERS.

Additional color graphics may be available in the e-book version of this book.

LIBRARY OF CONGRESS CATALOGING-IN-PUBLICATION DATA

BIOPOL 2009 (2009)
  Biodegradable polymers and sustainable polymers (BIOPOL-2009) / [edited by] Alfonso Jiminez, G.E. Zaikov.
       p. cm.
  Selected papers from BIOPOL 2009.
  Includes index.
  ISBN 978-1-61209-520-2 (hardcover)
  1. Biodegradable plastics--Congresses. I. Jiminez, Alfonso, 1965- II. Zaikov, G. E. (Gennadii Efremovich), 1935- III. Title.
  TP1180.B55B567 2009
  620.1'92323--dc22
                            2010049659

Published by Nova Science Publishers, Inc. † New York

This volume is dedicated to the memory of Frank Columbus

On December 1st 2010, Frank H. Columbus Jr. (President and Editor-in-Chief of Nova Science Publishers, New York) passed away suddenly at his home in New York.

We lost our colleague, our good friend, a nearly perfect person who helped scientists from all over the world. Particularly Frank did much for the popularization of Russian and Georgian scientific research, publishing a few thousand books based on the research of Soviet (Russian, Georgian, Ukranian etc.) scientists.

Frank was born on February 26th 1941 in Pennsylvania. He joined the army upon graduation of high school and went on to complete his education at the University of Maryland and at George Washington University. In 1969, he became the Vice-President of Cambridge Scientific. In 1975, he was invited to work for Plenum Publishing where he was the Vice-President until 1985, when he founded Nova Science Publishers, Inc.

Frank Columbus did a lot for the prosperity of many Soviet (Russian, Georgian, Ukranian, Armenian, Kazakh, Kyrgiz, etc.) scientists publishing books with achievements of their research. He did the same for scientists from East Europe – Poland, Hungary, Czeckoslovakia (today it is Czeck republic and Slovakia), Romania and Bulgaria.

He was a unique person who enjoyed studying throughout the course of his life, who felt at home in his country which he loved and was proud of, as well as in Russia and Georgia.

There is a famous Russian proverb: "The man is alive if people remember him." In this case, Frank is alive and will always be in our memories while we are living. He will be remembered for his talent, professionalism, brilliant ideas and above all – for his heart.

Contents

Preface		xi
Chapter 1	Biodegradation and Medical Application of Microbial Poly(3-hydroxybutyrate) *A. P. Bonartsev, A. L. Iordanskii, G. A. Bonartseva and G. E. Zaikov*	1
Chapter 2	Use of Hydroxytyrosol as Polypropylene Stabilizer and as a Potential Active Antioxidant *M. Peltzer, N. López, L. Matisová-Rychlá, J. Rychlý and A. Jiménez*	37
Chapter 3	Mechanical Properties of Dimer Fatty Acid-based Polyamides Biocomposites *Elodie Hablot, Rodrigue Matadi, Said Ahzi and Luc Avérous*	51
Chapter 4	Preparation and Properties of Three Layer Sheets Based on Gelatin and Poly(Lactic Acid) *J. F. Martucci and R. A. Ruseckaite*	67
Chapter 5	Evaluation of the Use of Natural Plasticizers in Commercial Lids for Food Packaging. Characterization and Migration in Food Simulants *C. Bueno-Ferrer, M. C. Garrigós and A. Jiménez*	83

Chapter 6	Evaluation of Parameters Essential for Efficiency in the Composting Process J. Klein, M. Zeni, V. T. Cardoso, B. C. D. A. Zoppas, A. M. C. Grisa and R. N. Brandalise	93
Chapter 7	Characterization of PP Films with Carvacrol and Thymol as Active Additives M. Ramos, M. A. Peltzer and M. C. Garrigós	105
Chapter 8	Lipase Catalyzed Synthesis of Biopolyester and Related Clay-based Nanohybrids Hale Öztürk, Eric Pollet, Anne Hébraud and Luc Avérous	117
Chapter 9	Characterization and Thermal Stability of Almonds by the Use of Thermal Analysis Techniques Arantzazu Valdés-García, Ana Beltrán-Sanahuja and M. Carmen Garrigós-Selva	137
Chapter 10	Chitosan as an Antimicrobial Agent for Footwear Leather Components M. C. Barros, I. P. Fernandes, V. Pinto, M. J. Ferreira, M. F. Barreiro and J. S. Amaral	151
Chapter 11	Characterization of Lignocellulosic Materials by Morphological and Thermal Techniques M. I. Rico, M. C. Garrigós, F. Parres and J. López	167
Chapter 12	Effect of Processing Methods on Mechanical Properties of Soya Protein Films P. Guerrero, L. Martin, S. Cabezudo and K. de la Caba	179
Chapter 13	Development of a Biodegradability Evaluation Method for Leather Used in the Footwear Industry M. A. de la Casa-Lillo, A. Diaz-Tahoces, P. N. de Aza-Moya, P. Mazón-Canales, V. Segarra-Orero, M. A. Martínez-Sanchez and M. Bertazzo	191

Chapter 14	Chromium Tanned Leather Waste Acid Extraction, Residue Recycling and Anaerobic Biodegradation Tests on Extracts *Maria J. Ferreira, Manuel F. Almeida, Vera Pinto, Isabel Santos, José L. Rodrigues, Fernanda Freitas and Sílvia Pinho*	**205**
Chapter 15	How to Shift Toughness of PLA into Non-break Area and to Create High Impact Flax Fibre Reinforcements *R. Forstne and W. Stadlbauer*	**225**
Index		**237**

PREFACE

Mankind currently produces polymers in amounts similar to cast iron, steel, and color metals all together if calculating production of polymers not by weight, but by volume. One can say, that mankind currently lives in a Polymer Age. Special emphasis should be given to biopolymers since they can offer properties similar to those of oil-based polymers with the added value of their extraordinary ability to be introduced in biological cycles with production of new raw materials for new polymers. This new book presents current research in the area of bio-based polymers and sustainable composites.

Chapter 1 - This review is designed to be a comprehensive source for a biodegradable polymer, poly(3-hydroxybutyrate), research including fundamental structure/properties relationships and biodegradation kinetics for samples of different morphology. In addition, this review focuses on applications of PHB in biomedicine and environment with discussion on commercial applications and health/safety concerns for biodegradable materials.

Chapter 2 - The use of natural antioxidants as stabilizers in polymer formulations has raised some interest in recent years making attractive their use in packaging applications. One of the main sources of such phenolic compounds is olive fruit which contains a wide variety of bioactive components. Among them, hydroxytyrosol (HT) stands out as a high value compound due to its antioxidant character and beneficial properties to food and human health. The aim of this study is to characterize and evaluate the stabilizing performance of HT in polypropylene (PP) films as a potential polymer stabilizer and active additive. Oxidation Induction Time (OIT) of selected samples of polypropylene (PP) additivated with HT was determined by Differential Scanning Calorimetry (DSC) and Chemiluminescence (CL). In addition, the Onset Oxidation Temperature (OOT) and the apparent activation energy (E_a) were determined by DSC and thermogravimetric analysis (TGA), respectively. HT showed good performance as PP stabilizer at concentrations higher than 0.05 wt%, since it was observed an increment in the OIT, OOT and E_a values. HT at concentrations higher than 0.5 wt% could be considered as active additive in PP films.

Chapter 3 - Nowadays, replacing petroleum-based materials with renewable resources is a major concern in terms of both economical and environmental points of view. In good agreement with this emergent concept of sustainable development, this work deals with the study of innovative green polyamides (DAPA) based on rapeseed oil-dimer fatty acid reinforced with cellulose fibres (CF). Mechanical properties of DAPA and its biocomposites were examined. Tensile tests were used to follow the effect of strain rate, temperature and filler content on the Young modulus and the yield stress. Tensile tests revealed that the Young modulus and the yield stress highly increased with increasing CF concentration. Halpin-Tsai and Eyring's micro-mechanical models were found to successfully predict the Young modulus and the yield stress respectively, of DAPA and DAPA/Cellulose biocomposites. These results clearly highlighted the attractive properties of DAPA-based biocomposites and demonstrated that these materials are good alternatives to conventional composites.

Chapter 4 - Glycerol plasticized-gelatin (Ge-30Gly) and poly(lactic acid) (PLA) films were prepared by heat-compression molding and then piled together to produce a biodegradable three-layer sheet with PLA as both outer layers and Ge-30Gly as inner layer. Multilayer sheets displayed a compact and uniform microstructure due to the highly compatible individual layers which could interact by strong hydrogen bonding. Lamination reduced total soluble matter compared to the single layers while keeping transparency. Tensile strength of the multilayer sheet (36.2 ±4.3 MPa) increased 16-folds compared with that of Ge-30Gly. Lamination also had beneficial effect on the barrier properties. The Water Vapour Permeation of the multilayer sheet ($1.2 \pm 0.1 \ 10^{-14}$ $kg \cdot m \cdot Pa^{-1} \cdot s^{-1} \cdot m^{-2}$) decreased compared with that of Ge-30Gly while the oxygen permeability (17.1 ± 2.3 $cm^3(O2) \cdot mm \cdot m^{-2} \cdot day^{-1}$) was reduced with respect to that of neat PLA and was comparable to the value of Ge-30Gly layer. The individual layers and multilayer sheet were submitted to degradation under indoor soil burial conditions. Biodegradation of multilayer as well as individual components was evaluated by monitoring water absorption and weight loss. During the experiment time, weight loss of multilayer sheet showed two stages. During the first stage, the weight loss of the film increased rapidly due to the loss of Ge-30Gly inner layer. From 25th to 120th day no further changes were observed suggesting that the inner layer was completely biodegraded. Visually, multilayer sheet turned from transparent to opaque after 120 days due to an increment of the crystallinity of the PLA polymer matrix. SEM analysis showed that after 25 days a significant change in the overall morphology of multilayer sheet could be observed. This result suggested that Ge inner layer promote the PLA outer layer degradation, probably because the Ge hydrophilicity allows the water molecules to penetrate more easily within the PLA and to trigger the hydrolytic degradation process faster.

Chapter 5 - Epoxidized soybean oil (ESBO) is a vegetable oil widely used as plasticizer and/or stabilizer for poly (vinyl chloride) (PVC) formulations in food contact materials, in particular in gaskets of lids for glass jars. ESBO shows some advantages, such as low toxicity to humans and biodegradable nature, compared to common additives for PVC, making very attractive its use in food packaging. However, its migration to foodstuff is a crucial issue for this application. The current specific migration limits (SME) established for ESBO in food contact materials make necessary the study of its presence and migration to food simulants, since the tolerable daily intake (TDI) of 1 mg kg^{-1} body weight is often exceeded.. In this work, a wide screening of the use of ESBO in commercial lids for glass jars was carried out and this additive was identified and determined in most of the commercial samples tested. ESBO migration to food simulants was also determined for controlling compliance of the current legal migration limits. Thermal characterization of commercial lids was also carried out in order to explain differences in composition.

Chapter 6 - The degradation of oxo-biodegradable polymeric materials after their use can be evaluated when they are exposed to different mediums, such as landfill, compostage; in direct contact with selected colonies of microorganisms, and others. The biological environment in which the polymers are generally disposed includes the presence of microorganisms. They use various sources of food and polymers may provide important substrates to obtain energy and produce new cells. This study proposes to look at the evolution of the degradation of organic matter by compostage, with PE films containing prooxidant additives (1 and 1,5 wt%), by evaluation of some parameters such as pH, humidity, volatile solids, organic carbon, total nitrogen and temperature, as well as the fungi and metals characterization.

Chapter 7 - Polypropylene (PP) active films were prepared and characterized by incorporating thymol and carvacrol as active additives. Different concentrations were studied: 4, 6 and 8 wt% of carvacrol, thymol, and carvacrol and thymol (1:1) mixtures. A PP film without any active compound was also prepared as control. A complete characterization of all formulations was performed by testing different thermal, functional, structural and mechanical properties. TGA results showed that the addition of the studied additives did not affect the PP thermal stability. DSC confirmed the stabilization against thermo-oxidative degradation, with higher oxygen induction parameters obtained for materials with additives. Neat PP showed lower oxygen transmission rate (OTR) values than the active films. SEM images showed certain porosity for films with higher concentrations of thymol or carvacrol. Tensile tests showed important differences between pure PP and formulations containing additives, in particular in elongation at yield and elastic modulus values, which could be due to a certain plasticization effect. The obtained results showed that these additives partially remained in the polymer matrix after processing and consequently they could be released during their shelf-life, acting as active additives in PP formulations.

Chapter 8 - Biosynthetic pathway, like enzymatic ring opening polymerization (ROP) of lactones, attracts attention as a new trend of biodegradable polymer synthesis due to its non-toxicity, mild reaction requirement and recyclability of immobilized enzyme. Besides the enzyme-catalyzed synthesis of biopolyesters, key researches are conducted nowadays on the elaboration of biocomposites in combination with inorganic (nano)particles. The goal is to improve some of these polyesters properties for specific biomedical applications. In parallel, the use of clays as inorganic porous supports to immobilize enzymes has also been described. This chapter aims at presenting the use and development of original catalytic systems based on lipases which are efficient for polyester synthesis and allowing the preparation of hybrid materials based on clay nanoparticles grafted with such polyesters. For this, ε-caprolactone (ε-CL) polymerization catalyzed by Candida antarctica lipase B (CALB) was carried out in the presence of montmorillonite and sepiolite clays to obtain organic/inorganic nanohybrids through polymer chains grafting and growth from the hydroxyl groups of the clay. Both the free form and immobilized form of CALB have been tested as catalytic systems and their efficiency has been compared. The polymerization kinetics and resulting products were fully characterized with NMR, SEC, DSC and TGA analyses.

Chapter 9 - Nuts are subjected to thermal processes in the elaboration of manufactured products that can affect their thermal stability and lead to oxidation processes. In this paper, almond samples from three different cultivars (Spanish Guara; Marcona and Butte from USA) have been characterized by the use of Differential Scanning Calorimetry (DSC). Thermal stability of samples has also been evaluated by the use of Thermogravimetric Analysis (TGA). Fruit and oil samples have been studied at different scanning rates. Clear differences between Spanish and American almonds were observed by applying multivariate statistical analysis to DSC and TGA results, proving the suitability of the proposed methods for discrimination between different almond cultivars. Fruit and oil samples can be used for this purpose, although the use of fruit samples has the advantage of shorter sample preparation times.

Chapter 10 - Chitosan is being increasingly used in distinct areas such as pharmaceutical, biomedical, cosmetics, food processing and agriculture. Among the interesting biological activities that have been ascribed to chitosan, the antimicrobial activity is probably the one to generate the higher number of applications. Within this work the role of chitosan in diverse applications has been reviewed with particular emphasis for those exploring its antimicrobial power. Furthermore, the mechanism to explain the antimicrobial activity of this emerging biopolymer is also discussed. The viability of using chitosan to effectively provide a functional coating for leather products was presented through an experimental case study. Results confirmed the potential of using this strategy to create antimicrobial leather products to be used, e.g., in the footwear industry.

Chapter 11 - The search for new materials is nowadays being focused on the use of biodegradable polymers such as lignocellulosic by-products. In this paper, different biomass products as nut shells and wood have been characterized in order to study their morphological and thermal behaviour. Scanning electron microscopy (SEM) has been used for the surface morphological study of different nut shells and wood. Structures of both samples have been compared in terms of possible expected mechanical properties. Sequential pyrolysis-gas chromatography-mass spectrometry (Py-GC-MS) was used for the thermal characterization of almond shells, where some characteristic volatile compounds were obtained with a decrease as sequential cycles were applied. The obtained results provided useful information on the thermal degradation of cellulose and lignin obtained from almond shells.

Chapter 12 - Glycerol-plasticized soya protein films were prepared through three different methods: solvent casting, compression, and freeze-drying followed by compression. The effects of processing method and glycerol content on the mechanical properties of soya protein films were studied. Young's modulus, tensile strength and elongation at break were evaluated and related to these two variables: processing and plasticizer amount. Results have been explained by using Fourier transform infrared (FTIR) spectroscopy.

Chapter 13 - This article aims to contribute to find of a solution to one of the main environmental problems in the footwear industry: the large quantity of waste which is produced and the long life-cycle of the materials due to the use of chromium compounds in tanning processes. Alternative tanning methods are being studied which allow a faster degradation rate of the materials used. In ordert to face this situation, it is first necessary to develop a quick method for the determination of the biodegradability to be able to foresee which of these products would be more biodegradable. INESCOP, together with the Miguel Hernández University of Elche (UMH), is developing a method based on the principle of natural biodegradation of organic compounds by microorganisms' action. The final objective is to get a method for a quick measurement of biodegradability of leather tanned with different industrial tanning methods, and to be able to establish which process should be used in order to achieve a lower environmental impact.

Chapter 14 - Chromium sulfate tanned leather wastes are currently mainly landfilled, despite their content in valuable biopolymers and minerals. A more sustainable option might be recovering as much as possible of its chromium and, consequently, lowering its content in the resulting leather scrap, thus facilitating the recycling of the remaining material. With this objective, chromium leather scrap was leached with sulfuric acid solutions at low temperature (296 and 313 K) aiming at maximizing chromium removal with minimum attack to the leather matrix. Although reasonable Cr recoveries were achieved (30 and 60 %), chromium in eluate from leaching the residue according to DIN 38414-S4:1984 standard exceeded the threshold value, being considered hazardous. Thus, it was methodically washed with water and alkaline solutions in order to remove or stabilize the chromium de-linked from collagen. Furthermore, the so-treated leather scrap was recycled in rubber compounds for shoe soles application, confirming its potential to improve tear resistance. However, fine tune of the formulation is necessary to avoid failure due to poor tensile strength. The anaerobic biodegradation of the acid solution resulting from Cr recovery was evaluated indicating anaerobic biodegradability in the range of 25 to 45 %.

Chapter 15 - PLA compounds with various impact modifiers were tested and improvements in Charpy impact strength up to factors of ten were achieved with new Core-Shell type modifiers opening the non-break area for PLA. Reinforcements with flax fibres increased tensile properties and dimensional stability on costs of toughness.

In: Biodegradable Polymers ...
Editors: A. Jimenez and G. E. Zaikov

ISBN 978-1-61209-520-2
© 2011 Nova Science Publishers, Inc.

Chapter 1

BIODEGRADATION AND MEDICAL APPLICATION OF MICROBIAL POLY(3-HYDROXYBUTYRATE)

A. P. Bonartsev[1,3,*], A. L. Iordanskii[2,†], G. A. Bonartseva[1] and G. E. Zaikov[2]

[1]A. N. Bach's Institute of Biochemistry, Russian Academy of Sciences, Leninskii prosp. 33, 119071 Moscow, Russia
[2]General Institute of Chemical Physics, Russian Academy of Sciences, Kosygin str. 4, 119991 Moscow, Russia
[3]Faculty of Biology, Moscow State University, Leninskie gory 1-12, 119992 Moscow, Russia

ABSTRACT

This review is designed to be a comprehensive source for a biodegradable polymer, poly(3-hydroxybutyrate), research including fundamental structure/properties relationships and biodegradation kinetics for samples of different morphology. In addition, this review focuses on applications of PHB in biomedicine and environment with

---

[*] E-mail: bonar@inbi.ras.ru; ant_bonar@mail.ru
[†] E-mail: ioran@chph.ras.ru

discussion on commercial applications and health/safety concerns for biodegradable materials.

Keywords: microbial poly(3-hydroxybutyrate), non-enzymatic hydrolysis, enzymatic degradation, biodegradation in soil, degradation in animal tissues, biocompatibility

INTRODUCTION

Over the last decade an intense development of biomedical application of microbial poly((R)-3-hydroxybutyrate) (PHB) in producing biodegradable polymer implants and controlled drug release systems required comprehensive understanding of the PHB biodegradation process [1-3]. Examination of PHB degradation is also necessary for development of novel environmentally-friendly polymer packaging systems [4-6]. It is generally accepted that biodegradation of PHB both in living systems and in the environment occurs via enzymatic and non-enzymatic processes that take place simultaneously under natural conditions.[1,7]. Opposite to PGA and PLGA, PHB is considered to be moderately resistant to *in vitro* degradation as well as to biodegradation in living bodies. The rates of degradation are influenced by the characteristics of the polymer, such as chemical composition, crystallinity, morphology and molecular weight [8,9]. However, *in vitro* and *in vivo* applications have been intensively investigated. But available data are often incomplete and even contradictory. The presence of conflicting data can be partially explained by the fact that biotechnologically-produced PHB with standardized properties is relatively rare and it is not readily available due to a wide variety of PHB biosynthesis sources and different manufacturing processes.

Most of the papers used in this review investigated PHB degradation process for development of specific medical devices. Depending on the applied biomedical purposes, biodegradation of PHB was investigated under different morphologies: films and plates with various thickness [10-13], cylinders [13-16], mono-filament threads [17-18] and microspheres [19]. PHB was used from various sources, with different molecular weight and crystallinity. Besides, different technologies of PHB devices manufacturing affect such important characteristics as well as polymer porosity and surface structure [11,12]. Reports regarding the complex theoretical research of hydrolysis mechanisms, enzymatic degradation and *in vivo* biodegradation of

PHB are relatively rare [10-11,13,20-22]. Nevertheless, the effect of thickness, size and geometry of PHB devices, molecular weight and crystallinity on the mechanism of PHB hydrolysis and biodegradation was not yet well clarified.

NON-ENZYMATIC HYDROLYSIS OF PHB

Examination of *in vitro* hydrolytic degradation of natural poly((R)-3-hydroxybutyrate) is a very important step for understanding biodegradation. Some careful studies on PHB hydrolysis were carried out 10-15 years ago [20-23] under standard experimental conditions simulating internal body fluids: i.e. buffered solutions with pH = 7.4 at 37 °C, but in some cases higher temperature (55 °C, 70 °C) and other pH values (from 2 to 11) were selected.

Classical experiments for PHB hydrolysis in comparison with other widespread biopolymer, polylactic acid (PLA), was carried out by Koyama and Doi [20]. They prepared PHB (10 x 10 mm^2 size, 50 µm thickness, 5 mg initial mass, M_n = 300 kDa, M_w = 650 kDa) and PLA films (M_n = 9 kDa, M_w = 21 kDa) by solvent casting and aged for 3 weeks to reach equilibrium. They showed that hydrolytic degradation of natural PHB is a very slow process. PHB films remained unchanged at 37 °C in 10 mM phosphate buffer (pH = 7.4) over a period of 150 days, while PLA films rapidly changed with time and reached 17% of the initial mass after 140 days. The decrease rate in PHB M_n was also much slower than PLA M_n since PHB decreased approximately to 65% after 150 days, while PLA decreased to 20% (2 kDa) of the initial PLA M_n at the same time point. We examined rates of *in vitro* mass loss of polymer films with the same thickness (40 µm) from PLA and PHB with the same molecular weight (M_n = 450 kDa). It was shown that the mass of PLA film decreased to 87%, whereas the mass of PHB film remained unchanged at 37 °C in 25 mM phosphate buffer (pH=7.4) over a period of 84 days [24-25].

The cleavage of polyester chains is known to be catalyzed by the carboxyl end groups, and its rate is proportional to water and ester concentrations that may be constant during the hydrolysis, owing to the presence of a large excess of water molecules and of ester bonds of polymer chains. Thus, the kinetics of non-enzymatic hydrolysis can be expressed by the following equation [26-27]:

$$\ln M_n = \ln M_n^0 - kt \qquad (1)$$

where M_n and M_n^0 are the number-average molecular weights of a polymer component at time t and zero, respectively.

The average number of bond cleavage per polymer molecule, N, is given by equation 2:

$$N = (M_n^0/M_n) - 1 = k_d P_n^0 t \qquad (2)$$

where k_d is the rate constant of hydrolytic degradation and P_n^0 is the number-average degree of polymerization at zero time. Thus, if the chain scission is completely random, the value of N is linearly dependent on time.

The decrease in molecular weight with time is the distinguishing feature of degradation mechanism in non-enzymatic hydrolysis in contrast to enzymatic hydrolysis of PHB where M_n values remained almost unchanged. It was supposed also that water-soluble oligomers of PHB may accelerate chain scissions of the polymer [20]. In contrast, Freier et al. [11] showed that PHB hydrolysis was not accelerated by the addition of pre-degraded PHB. The rate of mass and M_w loss in blends (70/30) from high (M_w = 641 kDa) and low-molecular weight PHB (M_w = 3 kDa) was the same. Meanwhile, the addition of amorphous atactic PHB (atPHB) (M_w = 10 kDa) to blends with high-molecular weight PHB caused significant acceleration of hydrolysis, i.e. the mass loss of PHB/atPHB blends was 7% in comparison with 0% mass loss of pure PHB, the decrease in M_w was 88% in comparison with 48% M_w decrease of pure PHB [11,28]. We showed that the rate of hydrolysis of PHB films depends on M_w. Films from PHB of high molecular weight (450 and 1000 kDa) degraded slowly as it was described above whereas films from PHB of low molecular weight (150 and 300 kDa) lost weight gradually and more rapidly [24-25].

In order to enhance the hydrolysis of PHB a higher temperature was selected for degradation experiments [20]. It was shown that the films weight (12 mm diameter, 65 μm thick) from PHB (M_n = 768 and 22 kDa, M_w = 1460 and 75 kDa) were unchanged at 55 °C in 10 mM phosphate buffer (pH=7.4) over a period of 58 days. The M_n value decreased from 768 to 245 kDa for 48 days. The film thickness increased from 65 to 75 μm for 48 days, suggesting that water permeated the polymer matrix during the process. The examination of the surface and cross-section of PHB films before and after hydrolysis showed that surface after 48 days was apparently unchanged, while the cross-section of films exhibited a porous structure (pore size < 0.5 μm). It was also shown that the rate of hydrolytic degradation was not dependent on the PHB crystallinity. Data indicated that the non-enzymatic hydrolysis of PHB in water proceeds via a random bulk hydrolysis of ester bonds in the polymer chain and occurs throughout the whole film, since water fully permeates the polymer matrix [20-21]. Moreover, since the molecular weight distribution was unimodal over the whole process it was proposed a random chain scission both in the crystalline and the amorphous regions of PHB [11,29]. For synthetic amorphous atactic PHB it was shown that hydrolysis followed a two-step process. First, the random chain scission proceeds with a molecular weight decrease. Then, at a molecular weight of about 10000 Da, mass loss begins [23].

The analysis of literature data shows a great spread in values of PHB hydrolytic degradation rate. These differences can be explained by thickness of PHB films or morphology of PHB devices used for experiments as well as by their different sources, purity degree and molecular weight (Table 1). Some general conclusions could be drawn from Table 1. At acidic or alkaline aqueous media, PHB degrades more rapidly: 0% after 20 weeks (140 days) incubation in 0.01 NaOH (pH=11) (200 kDa PHB, 100 μm films) with surface changing [33], 0% after 180 days incubation of PHB threads in phosphate buffer (pH=5.2 and 5.9) [18], complete PHB films biodegradation after 19 days (pH=13) and 28 days (pH=10) [31]. It was demonstrated that after 20 weeks of exposure to NaOH solution, the surfaces of PHB films became rougher. From these results, it can be surmised that the non-enzymatic degradation of PHAs progresses on their surfaces before noticeable weight loss occurs (Figure 1) [33].

It was also shown that treatment of PHB films with 1M NaOH caused a reduction in pore size on film surfaces from 1-5 μm to around 1 μm indicating a partial surface degradation of PHB in alkaline media [34-35]. At higher temperatures no weight loss of PHB films was observed after 98 and 182 days incubation in phosphate buffer (pH = 7.2) at 55 °C and 70 °C, respectively [17], 12% and 39% of PHB (450 and 150 kDa, respectively) films after 84 days incubation at 70 °C [35,40], 50% and 25% after 150 days incubation of microspheres (250-850 μm diameter) from PHB (50 kDa and 600 kDa, respectively) [36].

Table 1. *In vitro* non-enzymatic hydrolysis of PHB

Device	Initial M_w kDa	Thickness μm	Conditions	Relative mass loss, %	Relative decrease of M_w, %	Time, days	Links
film	650	50	37°C, pH=7.4	0	35	150	20
film	640	100	37°C, pH=7.4	0	64	730	12
film	640	100	37°C, pH=7.4	0	45	364	11
film	450	40	37°C, pH=7.4	0	42	84	24-25
film	150	40	37°C, pH=7.4	12	63	84	24-25
film	279	-	37°C, pH=7.4	7.5	-	50	31
plate	-	500	37°C, pH=7.4	3	-	40	30
plate	380	1000	37°C, pH=7.4	0	-	28	37
plate	380	2000	37°C, pH=7.4	0	8	98	32
thread	470	30	37°C, pH=7.0	0	-	180	18
thread	-	-	37°C, pH=7.2	0	-	182	17
microspheres	50	250-850	37°C, pH=7.4	0	0	150	36
thread	470	30	37°C, pH=5.2	0	-	180	17
film	279	-	37°C, pH=10	100	-	28	31
film	279	-	37°C, pH=13	100	-	19	31
film	650	50	55°C, pH=7.4	0	68	150	20
plate	380	2000	55°C, pH=7.4	0	61	98	32
film	640	100	70°C, pH=7.4	-	55	28	11
film	150	40	70°C, pH=7.4	39	96	84	24-25
film	450	40	70°C, pH=7.4	12	92	84	24-25
microspheres	50	250-850	85°C, pH=7.4	50	68	150	36
microspheres	600	250-850	85°C, pH=7.4	25	-	150	36

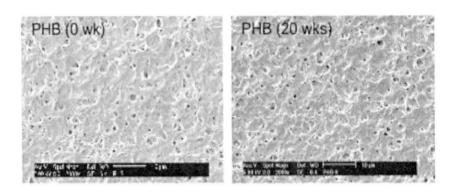

Figure. 1. Scanning electron microscopy photographs of PHB films both before (left) and after 20 weeks (right) of non-enzymatic hydrolysis in 0.01 N NaOH solution (scale bars, 10 μm) [33].

During *in vitro* degradation of PHB monofilament threads, films and plates a change in mechanical properties was observed under different conditions [17,37]. It was shown that these properties for threads became worse: load at break lost 36%, strain at break lost 33%, Young's modulus did not change, tensile strength lost 42% after 182 days incubation in phosphate buffer (pH = 7.2) at 70 °C. But at 37 °C changes were different, since at first load at break increased from 440 g to 510 g (16%) for 90 days and then it decreased to the initial value. Strain at break increased rapidly from 60 to 70% at 20th day and then gradually increased to 75% after 182 days. Young's modulus did not change [17]. PHB films showed a gradual 32% decrease and 77% fall in tensile strength during 120 days incubation in phosphate buffer (pH = 7.4) at 37 °C [37]. PHB plates showed more complicated changes. At first, tensile strength dropped 13% for 1 day and then increased to the initial value after 28 days, Young's modulus dropped 32% for 1 day and then remain unchanged, while stiffness decreased sharply 40% for 1 day and then remain unchanged [38].

ENZYMATIC DEGRADATION OF PHB IN VITRO

The examination of *in vitro* enzymatic degradation of PHB is another important step to understand the role of PHB in animal tissues and environment. Some authors reported PHB degradation by depolymerases [20-21]. At these early works it was shown that 68-85% mass loss of PHB (M_w = 650-768 and 22 kDa, respectively) films (50-65 μm thick) occurred for 20 h under incubation at 37 °C in phosphate solution (pH = 7.4) with depolymerase (1.5-3 μg/mL) isolated from *A. faecalis*. The rate of enzymatic degradation of PHB films (M_n = 768 and 22 kDa) was 0.17 and 0.15 mg/h, respectively. The thickness of polymer films dropped from 65 to 22 μm (32% of initial thickness) during incubation. SEM examination showed that the surface of the PHB film after enzymatic degradation was apparently blemished by the action of PHB depolymerase, while no change was observed in the bulk. Moreover, the molecular weight of PHB remained almost unchanged after enzymatic hydrolysis, decreasing from 768 to 669 kDa or even remaining unchanged (22 kDa) [20-21].

Data on enzymatic degradation of PHB by specific depolymerases was reported in detail [39]. It is important to note that PHB depolymerase is a very specific enzyme and the hydrolysis of polymer by depolymerase is a nonreversible process. But in animal tissues and even in the environment the enzymatic degradation of PHB occurs mainly by non-specific sterases [19,40].

The rate of enzymatic erosion of PHB by depolymerase is strongly dependent on the enzyme concentration. This is a heterogeneous reaction involving two steps, namely, adsorption and hydrolysis. The first step is the adsorption of the enzyme onto the PHB film surface by the binding domain of the PHB-depolymerase, while the second step is the hydrolysis of polyester chains by the active site of the enzyme. The rate of enzymatic erosion for chemosynthetic PHB samples containing both monomeric units of (R)- and (S)-3-hydroxybutyrate is strongly dependent on both, the stereocomposition and tacticity of samples. The water-soluble products of enzymatic PHB random hydrolysis showed a mixture of monomers and oligomers of (R)-3-hydroxybutirate. The rate of enzymatic hydrolysis for melt-crystallized PHB films by PHB-depolymerase decreased with an increase in the crystallinity of the PHB film, while the rate of enzymatic degradation for PHB chains in the amorphous state was approximately 20 times higher than the rate for PHB chains in a crystalline state. It was suggested that the PHB depolymerase predominantly hydrolyzes polymer chains in the amorphous phase and further erodes the crystalline phase. The surface of the PHB film after enzymatic degradation was apparently blemished by the action of PHB-depolymerase, while no change was observed inside the film. Thus, depolymerase hydrolyses of the polyester chains in the surface layer of the film proceeds in surface layers, while the PHB enzymatic degradation is affected by many factors as monomer composition, molecular weight and degree of crystallinity [39].

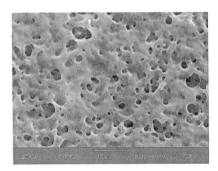 

Figure 2. Scanning electron microscopy photographs of (a) PHB film; (b) PHB film treated with lipase (0.1 g/l at 30°C and pH=7.0 for 24 h) [35].

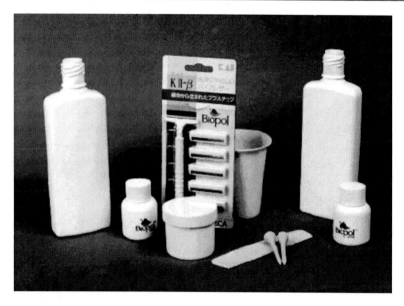

Figure 3. Molded PHB objects for various applications. In-soil burial or composting experiments [2].

It is also necessary to observe the PHB enzymatic degradation under conditions to model the animal tissues and body fluids containing non-specific esterases. *In vitro* PHB degradation in the presence of various lipases was carried out in buffered solutions containing lipases [41-42], in digestive juices (for example, pancreatin) [11], biological media (serum or blood) [18] and crude tissue extracts containing a mixture of enzymes [19]. It was noted that a Ser..His..Asp triad constitutes the active center of the catalytic domain of both PHB depolymerase [43] and lipases [44]. Serine is part of the pentapeptide Gly X1-Ser-X2-Gly, which has been located in PHB depolymerases as well as in lipases, esterases and serine proteases [43].

On the other hand, it was shown that PHB was not degraded during 100 days by lipases isolated from different bacteria and fungi [41,42]. The progressive PHB degradation by lipases was previously reported [24-25,34-35]. PHB enzymatic biodegradation was also studied in biological media. It was shown that pancreatin addition lead to no additional mass loss of PHB when compared with simple hydrolysis [11]. PHB degradation in serum and blood was demonstrated to be similar to hydrolysis processes in buffered solution [24-25], whereas progressive mass loss of PHB sutures was observed in serum and blood (16% and 25%, respectively after 180 days incubation) [18], while crude extracts from liver, muscles, kidney, heart and brain showed some activity to degrade PHB (from 2% to 18% mass loss after 17 h incubation at pH 7.5 and 9.5) [19]. The degradation rate in pancreatin solution was accelerated about threefold (34% decrease in M_w after incubation for 84 days in pancreatin vs. 11% decrease in M_w after incubation in phosphate buffer [11]. Similar data were obtained for PHB biodegradation in buffered solutions with porcine lipase addition (72% decrease in PHB M_w after incubation for 84 days with lipase (20 U/mg, 10 mg/ml in Tris-buffer) vs. 39% decrease in M_w after incubation in phosphate buffer) [24-25]. This observation is in contrast to enzymatic degradation by PHB-depolymerases which was reported to proceed on the polymer film surface with almost unchanged molecular weight [20-21]. It has been proposed that for depolymerases the enzyme relative size compared with the void space in solvent cast films is the limiting factor for diffusion into the polymer matrix whereas lipases can penetrate into the polymer matrix through pores in PHB film [34-35]. It was shown that lipase (0.1 g L^{-1} in buffer) caused significant morphological changes in PHB films surface by transferring from native PHB films with pores ranging from 1 to 5 μm into a pore-free surface. It was supposed that pores had a fairly large surface exposed to lipase, thus it was degraded more easily (Figure 2) [34-35]. Lipase can partially penetrate into pores of PHB film but the enzymatic degradation mainly proceeds on the surface of the polymer film. Two additional effects reported for depolymerases could be important. On one hand segmental mobility in amorphous phase and polymer hydrophobicity play an important role in enzymatic PHB degradation by non-specific esterases [11]. On the other hand, significant impairment on tensile strength was observed during PHB enzymatic biodegradation in serum and blood. It was shown that load at break lost 29%, Young's modulus lost 20%, and tensile strength did not change after 180 days of incubation [18].

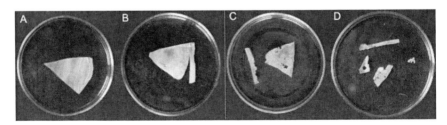

Figure 4. Undegraded PHB film (a) and PHB films with different degrees of degradation after 2 months incubation in soil suspension: anaerobic conditions without nitrate (b), microaerobic conditions without nitrate (c) and microaerobic conditions with nitrate (d) [24,48].

BIODEGRADATION OF PHB BY SOIL MICROORGANISMS

Polymers exposed to the environment are degraded through hydrolysis, mechanical, thermal, oxidative or photochemical biodegradation [4,32,45,46]. One of the valuable properties of PHB is biodegradability, which can be evaluated by using various laboratory tests. Requirements for PHB biodegradability may vary in accordance with its intended applications. PHB can be completely degraded by microorganisms getting finally CO_2 and H_2O. This property of PHB allows manufacturing biodegradable materials for various applications (Figure 3) [2].

The degradation of PHB and its composites in natural ecosystems, such as soil, compost and water, was described in some publications [2,32,45,46]. Maergaert et al. isolated more than 300 microbial strains from soil capable of degrading PHB [45]. Bacteria detected on the degraded PHB films were dominated by *Pseudomonas, Bacillus, Azospirillum, Mycobacterium,* and *Streptomyces*. PHB samples have been tested for fungicidity and resistance to fungi by estimating the growth rate of test fungi from *Aspergillus, Aureobasidium, Chaetomium, Paecilomyces, Penicillum, Trichoderma* under optimal growth conditions. PHB films did exhibit neither fungicide properties nor the resistance to fungal damage, and served as a good substrate for fungal growth [47].

Biodegradability of PHB films was studied under aerobic, microaerobic and anaerobic conditions in the presence and absence of nitrate by microbial populations of soil, sludge from anaerobic and nitrifying/denitrifying reactors (Figure 4) [48]. Changes in molecular weight, crystallinity, and mechanical properties of PHB were studied. A correlation between the PHB degradation degree and its molecular weight was demonstrated. The most degraded PHB exhibited the highest values of crystallinity. As it has been shown by Spyros et al., PHAs contain amorphous and crystalline regions, of which the former are much more susceptible to microbial attack [49]. The PHB microbial degradation must be associated with a decrease in its molecular weight and an increase in crystallinity. Moreover, microbial degradation of the amorphous regions of PHB films made them more rigid and the structure of the polymer much looser [48].

PHB biodegradation in the enriched culture obtained from soil on the medium used to cultivate denitrifying bacteria has been also studied. The dominant bacterial species, *Pseudomonas fluorescens* and *Pseudomonas stutzeri*, were identified in this culture. Under denitrifying conditions, PHB films were completely degraded for 7 days. Both, the film weight and M_w

decreased with time. In contrast to the data by Doi et al. [21] who found that M_w of PHB remained unchanged upon enzymatic biodegradation in an aqueous solution of PHB-depolymerase from *Alcaligenes faecalis*, in our experiments the average viscosity molecular weight decreased gradually from 1540 to 580 kDa and from 890 to 612 kDa, respectively. The "exo"-type cleavage of the polymer chain, i.e a successive removal of the terminal groups, is known to occur at a higher rate than the "endo"-type cleavage, i.e, a random breakage of the polymer chain at the enzyme-binding sites. Thus, the former type of polymer degradation is primarily responsible for changes in the average molecular weight. However the "endo"-type attack plays an important role at the initiation of biodegradation, because at the beginning, a few polymer chains are oriented and their ends are accessible to the enzyme [50]. Biodegradation of the lower-molecular polymer, with a higher number of terminal groups, is more active, probably, because the "exo"-type degradation is more active in lower than in higher molecular polymer [48,51].

BIODEGRADATION OF PHB IN VIVO IN ANIMAL TISSUES

The first scientific works on *in vivo* PHB biodegradation in animal tissues were carried out 15-20 years ago by Miller et al and Saito et al [17,19]. As it was noted above, both enzymatic and non-enzymatic processes of *in vivo* PHB biodegradation can occur simultaneously under normal conditions. But this fact does not mean that *in vivo* polymer biodegradation is a simple combination of non-enzymatic hydrolysis and enzymatic degradation. As it was noted above for *in vitro* PHB hydrolysis, the main reason for controversy is the use of samples obtained from various processing technologies as well as the incomparability of different implantation and animal models. Most researches on PHB biodegradation were carried out by using prototypes of various medical devices based on PHB: solid films and plates [10,13,24,52], porous patches [11,12], non-woven patches consisting on fibers [53-57], screws [24], cylinders as nerve guidance channels and conduits [13,15-16], monofilament sutures [17,18] and microspheres [19,58]. *In vivo* biodegradation researches were carried out on various laboratory animals differing in metabolism. The implantation of PHB devices on laboratory aimals was carried out through different ways: subcutaneously [10,13,17,18,24,59], intraperitoneally on a bowel [11], subperiostally on the osseus skull [12,52], nerve wrap-around [14-16], intramuscularly [58-59], into the pericardium [54-57], into the atrium [53] and intravenously [19]. The terms of implantation were also different and they are indicated in Table 2.

Table 2. *In vivo* PHB biodegradation

Device	Thickness µm	Surgical procedure	Relative mass loss of PHB, %	Relative loss of PHB molecular weight, %	Time, months	Links
film	1200	subcutaneously	1.6	43	6	13
film	150-200	Subcutaneously	6	60	6	10
film	50	Subcutaneously	100	100	3	24,25
porous PHB/ atactic PHB patch	100	intraperitoneally	>90*	62	6.5	11
porous PHB/ atactic PHB patch	250	with contact to bone	>50*	65	6.5	12
films and plates	100-1000	subperiostally	100	-	25	52
films and plates	100, 500	subperiostally	<10*	-	20	52
plates and screws	500, 1500	subperiostally	0	-	12	25
cylinder	150	nerve wrap-around	0	-	1	16
cylinder	150	nerve wrap-around	>25*	-	12	15
mono-filament suture	-	subcutaneously	0	-	6	17
mono-filament suture	30	subcutaneously	30	-	6	18
thin films	-	subcutaneously	>30*	-	2	59
non-woven patch	200-600	pericardium wall	>90*	-	24	55
non-woven patch	200-600	trans-annular patches	>99*	-	12	56
non-woven patch	200-600	septal of right atrium	>99*	-	12	53
non-woven patch	200-600	pericardium wall	27	-	24	57
microspheres	0.5-0.8	intravenously	8*	-	2	19
microspheres	100-300	intramuscularly (legs)	0*	-	2	58
rivet-shaped plate	2300	intraosseously	<10*	-	6	37

* Indirect data.

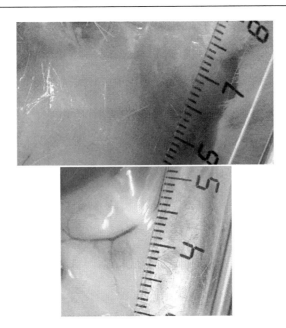

Figure 5. Biodegradation of PHB films *in vivo*. Connective-tissue capsule with PHB thin films (outlined with broken line) 2 weeks (98% residual weight of the film) (left); and 3 months (0% residual weight of the film) (righth) after subcutaneous implantation [24,25].

The most important study on *in vivo* PHB biodegradation was fulfilled by Gogolewski et al. and Qu et al. [10,13]. It was shown that PHB lost about 1.6% (injection-molded film, 1.2 mm thick, M_w = 130 kDa) [13] and 6% (solvent-casting film, 40 μm thick, M_w = 534 kDa) [10] of their initial weight after 6 months. The observed weight loss was partially due to leaching out of low molecular weight fractions and impurities initially present in the implants. PHB showed a constant increase in crystallinity (from 60.7 to 64.8%) up to 6 months [13] or an increase (from 65.0 to 67.9%) after 1 month with further drop to 64.5% after 6 months of implantation [10], suggesting that the degradation process had not affected the crystalline regions. These data are in accordance with those of PHB hydrolysis [20] and *in vitro* enzymatic PHB degradation by lipases [11] where M_w decreased. The initial biodegradation of the amorphous regions of PHB is similar to PHB degradation by depolymerase [39].

Thus, the observed biodegradation of PHB showed the coexistence of two different degradation mechanisms in polymer hydrolysis: enzymatically or non-enzymatically catalyzed degradation. Although non-enzymatic catalysis occurred randomly in the homopolymer, as it is indicated by the M_w loss rate, a critical molecular weight is reached some time when the enzyme-catalyzed hydrolysis accelerate degradation at the surface because the easier enzyme/polymer interaction. However when considering the PHB low weight loss, the critical molecular is not reached, resulting in low molecular weight and loss of crystallinity providing some sites for the enzymatic hydrolysis to accelerate PHB degradation [10,13]. Additional data revealing the mechanism of PHB biodegradation in animal tissues were obtained in long-term implantation experiments. A very slow degradation of films and plates was observed during 20 months. Some drop in PHB weight loss took place between the 20th and 25th month. Only initial signs of degradation were found on the implant surface after 20 months but no more test bodies could be detected after 25 months [52]. The complete *in vivo* biodegradation from 3 to 30 months of PHB was reported by other researchers [53,55-57,60], whereas almost no weight loss and surface changes in PHB during 6 months of *in vivo* biodegradation was shown [13,17]. Residual fragments of PHB implants were found after 30 months from the patches implantation [54,56]. A reduction in PHB patch size in 27% was shown in patients after 24 months after surgical procedure on pericardial closure [57]. It was shown that 30% mass loss in PHB sutures occurred gradually during 180 days with minor changes in the surface microstructure and in sutures volume [18]. It was shown that PHB non-woven patches were slowly degraded by polynucleated macrophages, and 12 months after surgery no PHB device was identifiable and only small particles were still observed. The absorption time of PHB patches was long enough to permit regeneration of a normal tissue [53]. The progressive biodegradation in PHB sheets was qualitatively demonstrated at 2, 6 and 12 months after implantation as weakening of the implant surface, tearing/cracking of implant, fragmentation and decrease in the polymer volume [15,37,59]. It was reported that the PHB biodegradation process consists of several phases. At the initial stage, PHB films were covered by fibrous capsules. At a second phase capsulated PHB films lost weight very slowly with simultaneous increase of crystallinity and decrease of M_w and mechanical properties. At third phase PHB films were rapidly disintegrated and then completely degraded. Finally, empty fibrous capsule were resolved (Figure 5) [24-25]. Interesting data were obtained for *in vivo* biodegradation of PHB microspheres (0.5-0.8 μm in diameter). It was indirectly demonstrated that PHB lost about 8 wt% of microspheres accumulated in liver after 2 month of intravenous injections. It was also demonstrated the presence of several types of PHB degrading enzymes in the animal tissues extracts [19].

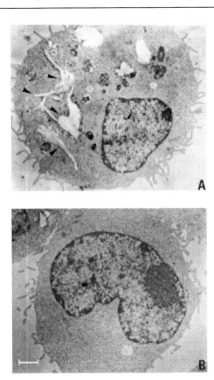

Figure 6. Phagocytosis of PHB microparticles in macrophages. TEM analysis of cultured macrophages in the presence (A) or absence (B) of 2 µg PHB microparticles mL^{-1} for 24 h. Bar in B represents 1 µm, for A and B.

Some researches studied a biodegradation of PHB threads by the determination of mechanical properties [17-18]. It was shown that at first, load at break index decreased rapidly from 440 g to 390 g (12%) at 15th day and then gradually increased to the initial value at 90th day and remain almost unchanged up to 182 days [17]; or gradually decreased 27% during 180 days [18]. Strain at break decreased rapidly from 60 to 50% at 10th day and then gradually increased to 70% after 182 days [17-18]. It was observed that the primary reason for *in vivo* PHB biodegradation was a lysosomal and phagocytic activity of polynucleated macrophages and giant cells of foreign body. The activity of tissue macrophages and non-specific enzymes in body liquids made a main contribution to increase significantly rate of *in vivo* PHB biodegradation compared with rate of *in vitro* PHB hydrolysis. PHB was encapsulated by degrading macrophages and stimulating the uniform macrophage infiltration, which is important not only for the degradation process but also for the restoration of functional tissues. The long absorption time produced a foreign-body reaction, which was restricted to macrophages forming layers [18,53,56,59]. It was observed a significant increase in activity of two specific lipases after 7 and 14 days of PHB contact with animal tissues. Moreover, liver specific genes were induced with similar results. It is striking that pancreatic enzymes were induced in the gastric wall after contact with biomaterials [40]. Saito et al suggested the presence of at least two types of degradative enzymes in rat tissues: liver serine esterases with maximum activity in alkaline media (pH = 9.5) and kidney esterases with maximum activity in neutral media [19]. The mechanism of PHB biodegradation by macrophages was demonstrated at cultured macrophages incubated with particles of low-molecular weight PHB [61]. It was shown that macrophages and, to a lower level, fibroblasts have the ability to take up PHB particles (1-10 µm). At high concentrations of PHB particles (>10 µg/mL) the phagocytosis is accompanied by toxic effects and alteration of the functional status of the macrophages but not for the fibroblasts. This process is accompanied by cell damage and cell death. The elevated production of nitric oxide (NO) and tumor necrosis alfa factor (TNF-α) by activated macrophages was also observed. It was suggested that the cell damage and cell death might be due to phagocytosis of large amounts of PHB particles filling up the cells. It was also demonstrated that phagocytized PHB particles disappeared due to active PHB biodegradation process (Figure 6) [61].

PHB APPLICATIONS

1. Medical Devices Based on PHB and *In vivo* PHB Biocompatibility

The most significant area for PHB applications is the development of implanted medical devices for dental, cranio-maxillofacial, orthopedic, cardiovascular, hernioplastic and skin surgery. A number of potential medical devices based on PHB, such as bioresorbable surgical sutures [17-18,62-63], biodegradable screws and plates for cartilage and bone fixation [24,52], biodegradable membranes for periodontal treatment, surgical meshes with PHB coating for hernioplastic surgery [24], wound coverings [64], patches for repair of a bowel, pericardial and osseous defects [11-12,53-57], nerve guidance channels and conduits [15-16] etc. were developed (Figure 7).

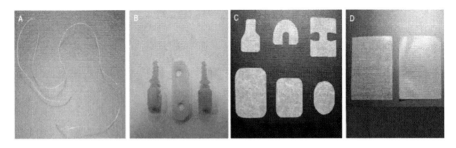

Figure 7. Medical devices based of PHB. (a) bioresorbable surgical suture; (b) biodegradable screws and plate for cartilage and bone fixation; (c) biodegradable membranes for periodontal treatment; (d) surgical meshes with PHB coating for hernioplastic surgery, pure (left) and loaded with antiplatelet drug, dipyridamole (right) [24].

Tissue reactions to implanted PHB films and medical devices were studied. In most cases a good biocompatibility of PHB was demonstrated. In general, no acute inflammation, abscess formation or tissue necrosis was observed in tissue surrounding of the implanted PHB materials. In addition, no tissue reactivity or cellular mobilization occurred in areas far away from the implantation site [10,13,24,58]. It was shown that PHB elicited similar mild tissue response, such as PLA did [13], but on the other hand the use of implants consisting of PLA, PGA and their copolymers provokes a number of consequences related with the chronic inflammatory reactions in tissues [65-69].

Subcutaneous implantation of PHB films for 1 month showed that samples were surrounded by a well-developed, homogeneous fibrous capsule about 80-100 μm in thickness. The vascularized capsule consisted primarily of connective tissue cells (mainly immature fibroblasts) aligned parallel to the implant surface. A mild inflammatory reaction was manifested by the presence of mononuclear macrophages, foreign body cells and lymphocytes. Three months after implantation, the fibrous capsules had thickened to 180-200 μm due to the increase in the amount of connective tissue cells and some collagen fiber deposits. A substantial decrease in inflammatory cells was observed after 3 months, tissues at the interface of the polymer were densely organized to form bundles. After 6 months of implantation, the number of inflammatory cells had decreased and the fibrous capsule, now thinned to about 80-100 μm, consisted mainly of collagen fibers and a significantly reduced amount of connective tissue cells. A little inflammatory cells effusion was observed in the tissue adherent to the implants after 3 and 6 months of implantation

[10,13]. Tissue reaction to films from PHB of different molecular weight (300, 450, 1000 kDa) implanted subcutaneously was relatively mild and did not change from tissue reaction to control glass plates [24].

The implantation of PHB with contact to bones lead to a favorable overall tissue response with high rate of early healing and new bone formation with indications of an osteogenic characteristic for PHB compared with other thermoplastics, such as polyethylene. Firstly, there was a mixture of soft tissues, containing active fibroblasts and rather loosely woven osteonal bones seen within 100 μm of the interface. There was no evidence of giant cells response within the soft tissue in the early stages of implantation. This tissue became more orientated in the direction parallel to the implant interface. The dependence of the bone growth on the polymer interface was demonstrated by the new bone growing away from the interface rather than towards it 3 months after implantation. During 6 months, the implant was closely encased in new bone of normal appearance with no interposed fibrous tissue. Thus, PHB-based materials produced superior bone healing [37].

Regeneration of neointima and neomedia, comparable to native arterial tissues, was observed at 3-24 months after implantation of PHB non-woven patches into the right ventricular outflow tract and pulmonary arteries. In the control group, a neointimal layer was observed but no neomedia comparable to native arterial tissue. Three layers were identified in the regenerated tissue: neointima with endothelium-like lining, neomedia with smooth muscle cells, collagenous and elastic tissue, and a layer with polynucleated macrophages surrounding PHB, capillaries and collagen tissue. It was concluded that PHB non-woven patches could be used as a scaffold for tissue regeneration in low-pressure systems. The regenerated vessel had structural and biochemical qualities in common with the native pulmonary artery [56]. Biodegradable PHB patches implanted in atrial septal defects promoted formation of regenerated tissues that resembled native wall. Regenerated tissues were found to be composed of three layers: a monolayer with endothelium-like cells, a layer with fibroblasts and some smooth-muscle cells, collagenous tissue and capillaries, and a third layer with phagocytizing cells isolating and degrading PHB. The neointima contained a complete endothelium-like layer resembling the native endothelial cells. The patch materials were encapsulated by degrading macrophages. There was a strict border between the collagenous and the phagocytizing layers. Presence of PHB seems to stimulate uniform macrophage infiltration, which was found to be important for the degradation process and the restoration of functional tissues. Lymphocytic infiltration as foreign-body reaction, which is common after the replacement of vessel walls with commercial woven Dacron patch, was wholly absent in this case. It was suggested that the absorption time of PHB patches was long enough to permit tissue regeneration strength enough to prevent development of shunts in the atrial septal position [53]. The prevention of post-operative pericardial adhesions by closure of the pericardium with absorbable PHB patch was demonstrated. The regeneration of mesothelial layer after implantation of PHB pericardial patchs was observed. The complete regeneration of mesothelium, with morphology and biochemical activity similar to native, may explain the reduction of post-operative pericardial adhesions after operations with insertion of absorbable PHB patches [55]. The regeneration of normal filament structure of restored tissues was observed by immuno-histochemical methods after PHB implantation [54]. The immuno-histochemical observation of cytokeratine, an intermediate filament which is constituent of epithelial and mesodermal cells, agreed with observations on intact mesothelium. Heparan sulfate proteoglycan, a marker of basement membrane, was also identified [54].

PHB patches for the gastrointestinal tract were tested using animal models. Patches made from PHB sutured and PHB membranes were implanted to close experimental defects of stomach and bowel wall. The complete regeneration of tissues was observed after 6 months after implantation without strong inflammatory response and fibrosis [11,70].

A recent application of biodegradable nerve guidance channels (conduits) for nerve repair procedures and nerve regeneration after spinal cord injury was reported. Tubular structures from PHB were modulated for this purpose. Successful nerve regeneration through a guidance channel was observed as early as after 1 month. Virtually all implanted conduits contained regenerated tissue cables centrally located within the channel lumen and composed of numerous myelinated axons and Schwann cells. The inflammatory reaction had not interfered with the nerve regeneration process. Progressive angiogenesis was present at the nerve ends and through the walls of the conduit. These results demonstrated good-quality nerve regeneration in PHB guidance channels [16,71].

Biocompatibility of PHB was evaluated by implanting microspheres (M_w = 450 kDa) into the femoral muscle of rats. They were surrounded by one or two layers of spindle cells and infiltration of inflammatory cells and mononuclear cells into these layers was recognized 1 week after implantation. After 4 weeks, the number of inflammatory cells had decreased and the layers of spindle cells had thickened. No inflammatory cells were observed after 8 weeks, while spheres were encapsulated by spindle cells. The toxicity of PHB microspheres was evaluated by weight change and survival times in L1210 tumor-bearing mice. No differences were observed compared with those of control. These results suggested that inflammation accompanying microsphere implantation is temporal with minimal toxicity to normal tissues [58].

The levels of tissue factors, inflammatory cytokines and metabolites of arachidonic acid were evaluated. Growth factors derived from endothelium and from macrophages were found. These factors probably stimulated both growth and regeneration when different biodegradable materials were used as grafts [40,53,55,70]. A positive reaction for thrombomodulin, a multi-functional protein with anticoagulant properties, was found in both mesothelial and endothelial cells after pericardial PHB patch implantation. Prostacycline production level, which was found to have cytoprotective effect on the pericardium, in the regenerated tissue was similar to that in native pericardium [53,55]. The PHB patch seemed to be highly biocompatible, since no signs of inflammation were macroscopically observed while the level of inflammation associated to cytokine m-RNA did not change, although a transient increase of interleukin-1β and interleukin-6 m-RNA through 1-7 days after patch implantation was detected. In contrast, tumor necrosis factor-α m-RNA was hardly detectable throughout the implantation period, which agreed well with a observed moderate fibrotic response [40,70].

2. PHB as Tissue Engineering Material and PHB *In Vitro* Biocompatibility

PHB is a promising material in tissue engineering due to high *in vitro* biocompatibility. Cell cultures of various sources, including murine and human fibroblasts [12,34,72-74], human mesenchymal stem cells [75], rabbit bone marrow cells (osteoblasts) [30,73,76], human osteogenic sarcoma cells [77], human epithelial cells [74,77], human endothelial cells [78-79], rabbit articular cartilage chondrocytes [80-81] and rabbit smooth muscle cells [82], in direct contact with PHB exhibited satisfactory levels of cell adhesion, viability and proliferation. It was also shown that fibroblasts, endothelium cells, and isolated hepatocytes cultured on PHB films exhibited high levels of cell adhesion and growth (Figure 8) [83].

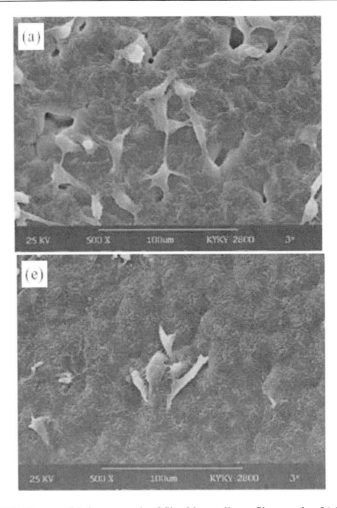

Figure 8. SEM image of 2 days growth of fibroblast cells on films made of (a) PHB; (e) PLA; (500 x) [73].

It was also shown that cultured cells produced collagen II and glycosaminoglycan, the specific structural biopolymers forming the extracellular matrix [77,80,81]. A good viability and proliferation level of macrophages and fibroblasts cell lines was obtained under culturing in presence of particles from short-chain low-molecular PHB [61]. However, it was shown that cell growth on PHB films was relatively poor: the viable cell number ranged from 1×10^3 to 2×10^5 [34,73,81]. An impaired interaction between PHB matrix and cytoskeleton of cultured cells was also demonstrated [77]. It was reported that a number of polymer properties including chemical composition, surface morphology, surface chemistry, surface energy and hydrophobicity play important roles in regulating cell viability and growth [84]. The investigation showed that this biomaterial could be used to make scaffolds for *in vitro* proliferous cells [34,76,80].

The most widespread methods to manufacture PHB scaffolds for tissue engineering by cell adhesion and growth on polymer surface are based on the modification of the PHB surface properties and microstructure by salt-leaching methods and enzymatic/chemical/physical treatment of polymer surface [34,76,80,85]. Adhesion to polymer substrates is one of the key issues in tissue engineering, because adhesive interactions control cell physiology. One of the most effective techniques to improve adhesion and cells growth on PHB films is by treatment of the polymer surface with enzymes, alkali or low pressure plasma [34,85]. Lipase treatment increases the viable cell number on the PHB film from 100 to 200 times when compared to the untreated PHB film. NaOH treatment on PHB film also indicated an increase in 25 times on the viable cell number compared with the untreated PHB film [34]. It was shown that treatment of PHB film surface with low pressure ammonia plasma improved growth of human fibroblasts and epithelial cells of respiratory mucosa due to increased hydrophilicity (but with no change of microstructure) in the polymer surface [74]. It was suggested that the improved hydrophilicity of films after treatment with lipases, alkali and plasma allowed cells to be easily attached on the polymer films compared to those on the untreated materials. The influence of hydrophilicity of the biomaterial on cell adhesion was demonstrated [86].

The PHB microstructure can be also responsible for cell adhesion and cell growth [87-89]. Therefore, modification of polymer film surface after enzymatic and chemical treatments (in particular, reduced pore size and surface smoothing) is expected to play an important role for enhanced cell growth on the polymer films [34]. For instance, osteoblasts preferred rougher surfaces with appropriate pore size [87,88] while fibroblasts prefer smoother surfaces [89]. Roughness affects cell attachment as it provides the right space for osteoblast growth or supplies solid anchors for filapodia. A scaffold with appropriate pore size provided better surface properties for anchoring type II collagen filaments and for their penetration into internal layers of the scaffolds implanted with chondrocytes. This could be caused by the interaction of extracellular matrix proteins with the material surface. The right surface properties may also promote cell attachment and proliferation by providing more space for gas/nutrients exchange or more serum protein adsorption. [30,76,80]. Sevastianov et al found that PHB films in contact with blood did not activate the hemostasis system at the level of cell response, but they activated the coagulation system [90].

The high biocompatibility of PHB may be due to several reasons. First of all, it is involved in important physiological functions for both, prokaryotes and eukaryotes. PHB from bacterial origin shows high stereospecificity that is inherent to biomolecules of all living bodies [91]. Low molecular weight PHB (up to 150 units of 3- D(-)-3-hydrohybutyric acid), complexed to other macromolecules (cPHB), was found to be a ubiquitous constituent of both prokaryotic and eukaryotic organisms of nearly all phyla [92-96]. cPHB was found in a wide variety of mammals tissues and organs: blood, kidney, vessels, nerves, vessels, eye, brain, as well as in organelles, membrane proteins, lipoproteins, and plaques. cPHB concentration ranged from 3-4 $\mu g\ g^{-1}$ wet tissue in nerves and brain to 12 $\mu g\ g^{-1}$ in blood plasma [97,98]. For humans, total plasma cPHB ranged from 0.60 to 18.2 mg L^{-1}, with an average value 3.5 mg L^{-1}. [98]. It was shown that cPHB is a functional part of ion channels for erythrocyte plasma membrane and hepatocyte mitochondria membrane [99,100]. The singular ability of cPHB to dissolve salts and to facilitate their transfer through hydrophobic barriers defines a potential physiological niche for cPHB in cell metabolism [94]. However the mechanism of PHB synthesis in eukaryotic organisms is not well clarified and requires additional studies. Nevertheless, it was suggested that cPHB is one of those products of symbiotic interaction between animals and microorganisms. It was shown, for example, that *E.coli* is able to synthesize low molecular weight PHB and cPHB plays various physiological roles in bacteria cell [96,101].

The intermediate product of PHB biodegradation, D(-)-3-hydroxybutyric acid, is also a normal constituent of blood at concentrations between 0.3 and 1.3 mM in all animal tissues [102,103]. PHB has a rather low degradation rate in the body in comparison to, e.g., PLGA, preventing the increase of 3-hydroxybutyric acid concentration in surrounding tissues [10,13], whereas PLA release, following local pH decrease in implantation area and acidic chronic irritation of surrounding tissues, is a serious problem in the application of medical devices on the base of PLGA [104,105]. Moreover, chronic inflammatory response to PLA and PGA that was observed in a number of cases may be induced by immune response to water-soluble oligomers released during their degradation [105-107].

3. Novel Drug Dosage Forms Based on PHB

An improvement in medical devices based on biopolymers to encapsulate different drugs opens up a wide field in applications in medicine. The design of injection systems for sustained drug delivery in the form of microparticles (microspheres, microcapsules) based on biodegradable polymers is extremely challenging in modern pharmacology. The fixation of pharmacologically active components by biopolymers and slow drug release from the microparticles provides an optimal level of drug concentration in local organs during long-term period (up to several months), providing effective pharmaceutical actions. At curative doses the prolonged delivery of drugs into organisms permits to eliminate the shortcomings in peroral, injectable, aerosol and the other traditional methods of drug administration. Among those shortcomings, hypertoxicity, instability, pulsative character of rate delivery and ineffective expenditure of drugs should be pointed out. Alternatively, applications of therapeutical polymer systems provide the deliverance for optimal doses which are very important in therapy of acute or chronic diseases [108]. An ideal biodegradable microsphere formulation would consist of free-flowing powder of uniform-sized microspheres lower than 125 μm in diameter with a high drug loading. The manufacturing method should produce such microspheres in a reproducible, scalable, and benign to some often delicate drugs process, with high encapsulation efficiency [109,110].

A number of drugs with various pharmacological activities were used for development of polymer controlled release systems on the base of poly(hydroxyl alcanoates), PHAs, mainly based on poly(3-hydroxybutyrate-co-3-hydroxyvalerate) and poly(3-hydroxybutyrate-co-4- hydroxybutyrate) copolymers: model drugs (2,7-dichlorofluorescein [111], dextran-FITC [112], methyl red [113,114], 7-hydroxethyltheophylline [115,116]); antibiotics and antibacterial drugs (rifampicin [117,118], tetracycline [119], cefoperazone and gentamicin [120], sulperazone and duocid [121-124], sulbactam and cefoperazone [125]); anticancer drugs (5-fluorouracil [126], 2',3'-diacyl-5-fluoro-2'-deoxyuridine [58]); anti-inflammatory drugs (indomethacin [127]); analgesics (tramadol [128]) vasodilator and antithrombotic drugs (dipyridamole [24,127,129], nitric oxide donor [130,131]) were used. The biocompatibility and pharmacological activity of some of these systems was studied [24,58,117,123-125,128,131]. But only few drugs were used for controlled release systems based on PHB homopolymer: 7-hydroxethyltheophylline, methyl red, 2',3'-diacyl-5-fluoro-2'-deoxyuridine, rifampicin, tramadol, indomethacin and dipyridamole [58,113-118,127-131].

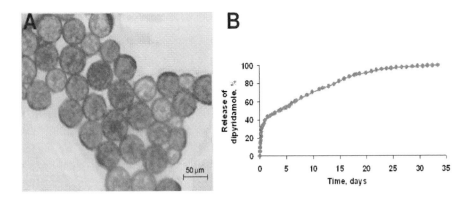

Figure 9. PHB microspheres for sustained delivery of drugs. (a) PHB microspheres (average diameter = 63 μm, PHB M_w = 1000 kDa) loaded with dipyridamole (10% w/w); (b) Sustained delivery of dipyridamole from PHB microspheres for more than 1 month [24,131].

The first drug-sustained delivery system based on PHB was developed by Korsatko et al., who observed a rapid release of encapsulated drug, 7-hydroxyethyltheophylline, from PHB tablets (M_w = 2000 kDa), as well as weight losses of PHB tablets containing the drug after subcutaneous implantation. It was suggested that PHB with molecular weight higher than 100 kDa was undesirable for long-term medication dosage [115].

Pouton and Akhtar described the release of low molecular weight drugs from PHB matrices and reported that the latter have a tendency for enhanced water penetration and pore formation [132]. The entrapment and release of the model drug, methyl red, from melt-crystallized PHB matrices was found to be a function of polymer crystallization kinetics and morphology whereas overall degree of crystallinity was shown to cause no effect on drug release kinetics. Methyl red released from PHB films for more than 7 days with initial rapid release ("burst effect") and a second phase with relatively uniform rate. Release profiles of PHB films crystallized at 110 °C exhibited a greater burst effect when compared to those crystallized at 60 °C. This was explained by better drug trapping within polymeric spherulites with the more rapid rates of PHB crystallization at 110 °C [113,114].

Kawaguchi et al showed that chemical properties of drugs and polymer molecular weight had an impact on drug delivery kinetics from PHB matrices [58]. Microspheres (100-300 μm in diameter) from PHB with different molecular weight (65, 135, and 450 kDa) were loaded with pro-drugs based on 5-fluoro-2'deoxyuridine (FUdR) synthesized by esterification with aliphatic acids (propionate, butyrate, and pentanoate). Pro-drugs had different physico-chemical properties, in particular, solubility in water (from 70 mg mL^{-1} for FUdR to 0.1 mg mL^{-1} for butyryl-FUdR). The release rates from the spheres depended on both the pro-drug lipophilicity and the polymer molecular weight. Regardless of the polymer, the relative release rates were propionyl-FUdR > butyryl- FUdR > pentanoyl-FUdR. The release of butyryl- FUdR and pentanoyl-FUdR from spheres consisting of low-molecular-weight PHB (M_w = 65 kDa) was faster than from the spheres of higher molecular weight (M_w = 135 or 450 kDa). The effect of drug content on the release rate was also studied. The higher the drug content in the PHB microspheres, the faster the drug release was. The release of FUdR continued for more than 5 days.

Kassab developed a well-managed technique for PHB microspheres manufacturing [118]. Microspheres were obtained within 5-100 μm size using a solvent evaporation method by changing the initial polymer/solvent ratio, emulsifier concentration, stirring rate and initial drug concentration. Very high drug loading of up to 408 g rifampicin/g PHB were achieved. Drug release rates were high: the maximal duration of rifampicin delivery was 5 days. Both the size and drug content of PHB microspheres were found to be effective in controlling the drug release from polymer microspheres.

The sustained release of analgesic drug, tramadol, from PHB microspheres was demonstrated by Salman et al [128]. It was shown that 58% of tramadol (the initial drug content in PHB matrix = 18%) was released from microspheres (7.5 μm in diameter) in the first 24 h. Drug release decreased with time. From 2 to 7 days the drug release showed zero-order rate. The entire amount of tramadol was released after 7 days.

Kinetics of different drug release from PHB films and microspheres was studied [24,127]. It was found that the release occurs by two mechanisms, diffusion and degradation, operating simultaneously. Vasodilator and antithrombotic drug, dipyridamole, and anti-inflammatory drug, indomethacin, showed that diffusion processes determined the release rate at the early stages of contact with the environment (the first 6-8 days). The coefficient of release diffusion of a drug depends on its nature, the thickness of PHB films containing the drug, the weight ratio of dipyridamole and indomethacin in the polymer and PHB molecular weight. A number of other drugs were also used for development of polymeric systems for sustained drug delivery: antibiotics (rifampicin, metronidazole, ciprofloxacin, levofloxacin), anti-inflammatory drugs (flurbiprofen, dexamethasone, prednisolone), and antitumor drugs (paclitaxel) [127]. Biodegradable microspheres based on PHB were designed for controlled release of dipyridamole. The release profiles for microspheres with different diameters 4, 9, 63, and 92 µm showed both a non-linear and linear behavior. Diffusion kinetic equations describing both linear (PHB hydrolysis) and non-linear (diffusion) stages of the dipyridamole release profiles from the spherical subjects were written down as the sum of two terms: desorption from homogeneous spheres in accordance with diffusion mechanisms and the zero-order release. In contrast to the diffusivity dependence on microsphere size, the constant characteristics of linearity were scarcely affected by the diameter of PHB microparticles. The kinetic profiles as well as the low rate of dipyridamole release were in agreement with kinetics of weight loss measured *in vitro* for PHB films and observed qualitatively for PHB microspheres. Taking into account those kinetic results, it was supposed that the degradation of both PHB films and microspheres is responsible for the linear stage of dipyridamole release profiles. The sustained invariable drug release is an essential property of injectable therapeutic polymer systems that allows keeping constant the adjusted drug dosing. PHB films and microspheres with sustained uniform drug release for more that 1 month showed that behavior (Figure 9) [24,127,129].

Biocompatibility and pharmacological activity of advanced drug delivery systems based on PHB was studied [24,58,117,128]. It was shown that implanted PHB films loaded with dipyridamole and indomethacin caused mild tissue reaction. The inflammation accompanying implantation of PHB matrices was temporary and toxicity relative to normal tissues was minimal [24]. No signs of toxicity were observed after administration of PHB microspheres loaded with tramadol [128]. A single intraperitoneal injection of PHB (M_w = 450 kDa) microspheres containing anticancer pro-drugs, butyryl-FUdR and pentanoyl-FUdR, resulted in high antitumor effects against P388 leukemia in mice over a period of 5 days [58]. Embolization with PHB microspheres *in vivo* to dogs was studied by Kasab et al. Renal angiograms obtained before and after embolization as well as histopathological observations showed the feasibility of using these microspheres as an alternative chemoembolization agent [117]. Epidural analgesic effects of tramadol released from PHB microspheres were observed for 21 h, whereas an equal dose of free tramadol was effective for less than 5 h. It was suggested that controlled release of tramadol from PHB microspheres is possible, and pain relief during epidural analgesia is prolonged by this drug formulation compared with free tramadol [128].

The observed data indicated the wide prospects in applications of drug-loaded medical devices and microspheres based on PHB as implantable and injectable therapeutic systems in medicine for treatment of various diseases:

cancer, cardio-vascular diseases, tuberculosis, osteomyelitis, arthritis etc. Besides application of PHB for producing medical devices and systems of sustained drug delivery, PHB can be used for production of systems for controlled release of enzymatic activators or inhibitors. The use of these systems allows the development of physiological models for *in vivo* prolonged local activation or inhibition of enzymes. PHB is a perspective tool in design of novel physiological models due to minimal adverse inflammatory tissue reaction to PHB matrices implantation or PHB microspheres administration. A system of sustained nitric oxide (NO) donor delivery based on PHB was developed. This system can be used for study of prolonged NO action on normal tissues of *in vivo* blood vessels. The development of *in vivo* models for prolonged NO local action on vascular tissues is a difficult problem, because NO donors deliver only for a few minutes. We developed a model of prolonged local NO action on appropriate artery based on PHB loaded with a new effective NO donor, FPTO [133]. It was shown that FPTO-loaded PHB cylinders could release FPTO (and consequently NO) for up to 1 month with relatively constant rate. FPTO-loaded PHB cylinders with sustained FPTO delivery were implanted around left carotid artery of Wistar rats, with pure PHB cylinders implanted around right carotid artery as control. After 1, 4 and 10 days after implantation, arteries and cylinders were isolated. The elevated levels of the main metabolic products of NO, nitrites and nitrates, in arterial tissues were observed indicating the possibility of application of this system for production of physiological models of NO prolonged action on *in vivo* arterial tissues [24,130,131].

ACKNOWLEDGMENTS

This review was supported by grants 06-04-49339 and 08-03-00929 from the Russian Foundation for Basic Research (RFBR) and by grant 07-II-10 from Foundation for Assistance to Small Innovative Enterprises (FASIE).

REFERENCES

[1] Chen GQ, Wu Q. *Biomaterials*, 2005; 26:6565-6578.
[2] Lenz RW, Marchessault RH. *Biomacromolecules*, 2005;6:1-8.
[3] Anderson AJ, Dawes EA *Microbiological Reviews*, 1990;54:450-472.

[4] Jendrossek D, Handrick R. *Annu Rev Microbiol.* 2002;56:403-432.
[5] Kim DY, Rhee YH. *Appl. Microbiol. Biotechnol.* 2003;61:300-308.
[6] Steinbuchel A, Lutke-Eversloh T. Biochem. Eng. J. 2003;16:81-96.
[7] Marois Y, Zhang Z, Vert M, Deng X, Lenz R, Guidoin R. *J. Biomater. Sci. Polym. Ed.*, 1999;10:483-499.
[8] Abe H, Doi Y. *Biomacromolecules.* 2002;3:133-138.
[9] Renstad R, Karlsson S, Albertsson AC. *Polym. Degrad. Stab.* 1999;63:201-211.
[10] Qu XH, Wu Q, Zhang KY, Chen GQ. *Biomaterials,* 2006;27:3540-3548.
[11] Freier T, Kunze C, Nischan C, Kramer S, Sternberg K, Sass M, Hopt UT, Schmitz KP. *Biomaterials,* 2002;23:2649-2657.
[12] Kunze C, Edgar Bernd H, Androsch R, Nischan C, Freier T, Kramer S, Kramp B, Schmitz KP. *Biomaterials,* 2006;27:192-201.
[13] Gogolewski S, Jovanovic M, Perren SM, Dillon JG, Hughes MK. *J. Biomed. Mater. Res.* 1993;27:1135-1148.
[14] Borkenhagen M, Stoll RC, Neuenschwander P, Suter UW, Aebischer P. *Biomaterials,* 1998;19:2155-2165.
[15] Hazari A, Johansson-Ruden G, Junemo-Bostrom K,. Ljungberg C, Terenghi G, Green C, Wiberg M. *J. Hand Surgery,* 1999;24B:291-295.
[16] Hazari A, Wiberg M, Johansson-Rudén G, Green C, Terenghi G. *British J. Plastic Surgery*, 1999;52:653-657.
[17] Miller ND, Williams DF. *Biomaterials*, 1987;8:129-137.
[18] Shishatskaya EI, Volova TG, Gordeev SA, Puzyr AP. *J Biomater Sci Polym Ed.* 2005;16:643-657.
[19] Saito T, Tomita K, Juni K, Ooba K. *Biomaterials,* 1991;12:309-312.
[20] Koyama N, Doi Y. *Can. J. Microbiol.*, 1995;41:316-322.
[21] Doi Y, Kanesawa Y, Kunioka M, Saito T. *Macromolecules*, 1990;23:26-31.
[22] Holland SJ, Jolly AM, Yasin M, Tighe BJ. *Biomaterials*, 1987;8:289-295.
[23] Kurcok P, Kowalczuk M, Adamus G, Jedlinrski Z, Lenz RW *JMS-Pure Appl. Chem.* 1995;32:875-880.
[24] Bonartsev AP, Myshkina VL, Nikolaeva DA, Furina EK, Makhina TA, Livshits VA, Boskhomdzhiev AP, Ivanov EA, Iordanskii AL, Bonartseva GA. *Current Research and Educational Topics and Trends in Applied Microbiology,* A. Méndez-Vilas (Ed), Formatex, Spain, 2007,295-307.
[25] Bonartsev AP. private communication.

[26] Cha Y, Pitt CG. *Biomaterials*, 1990;11:108-112.
[27] Schliecker G, Schmidt C, Fuchs S, Wombacher R, Kissel T. *Int. J. Pharm.* 2003;266:39-49.
[28] Scandola M, Focarete ML, Adamus G, Sikorska W, Baranowska I, Swierczek S, Gnatowski M, Kowalczuk M, Jedlinrski Z. *Macromolecules* 1997;30:2568-2574.
[29] Doi Y, Kanesawa Y, Kawaguchi Y, Kunioka M. *Makrom. Chem. Rapid. Commun.* 1989;10:227-230.
[30] Wang YW, Yang F, Wu Q, Cheng YC, Yu PH, Chen J, Chen GQ. *Biomaterials,* 2005;26:899-904.
[31] Muhamad II, Joon LK, Noor MAM. *Malaysian Polym. J.*, 2006;1:39-46.
[32] Mergaert J, Webb A, Anderson C, Wouters A, Swings *J. Appl. Environ. Microbiol.,* 1993;59:3233-3238.
[33] Choi GG, Kim HW, Rhee YH. *J. Microbiol.,* 2004;42:346-352.
[34] Yang X, Zhao K, Chen GQ. *Biomaterials* 2002;23:1391-1397.
[35] Zhao K, Yang X, Chen GQ, Chen JC. *J. Mat. Sci.: Mat. Medicine.* 2002;13:849-854.
[36] Wang HT, Palmer H, Linhardt RJ, Flanagan DR, Schmitt E. *Biomaterials,* 1990;11:679-685.
[37] Doyle C, Tanner ET Bonfield W. *Biomaterials*, 1991;12:841-847.
[38] Coskun S, Korkusuz F. Hasirci V. *J. Biomater. Sci. Polym. Ed.*, 2005;16:1485-1502.
[39] Sudesh K, Abe H, Doi Y. *Prog. Polym. Sci.* 2000;25:1503-1555.
[40] Lobler M, Sass M, Kunze C, Schmitz KP, Hopt UT. *Biomaterials*, 2002;23:577-583.
[41] Tokiwa Y, Suzuki T, Takeda K. *Agric. Biol. Chem.*1986;50:1323-1325.
[42] Hoshino A, Isono Y. *Biodegradation*, 2002;13:141-147.
[43] Jendrossek D, Schirmer A, Schlegel HG. *Appl. Microbiol. Biotechnol.* 1996;46:451-463.
[44] Winkler FK, D'Arcy A, Hunziker W. *Nature* 1990;343:771-774.
[45] Mergaert J, Anderson C, Wouters A, Swings J, Kersters K. *FEMS Microbiol. Rev.* 1992;9:317-321.
[46] Tokiwa Y, Calabia BP. *Biotechnol. Lett.* 2004;26:1181-1189.
[47] Mokeeva V, Chekunova L, Myshkina V, Nikolaeva D, Gerasin V, Bonartseva G. *Mikologia and Fitopatologia* 2002;36:59-63.
[48] Bonartseva GA, Myshkina VL, Nikolaeva DA, Kevbrina MV, Kallistova AY, Gerasin VA, Iordanskii AL, Nozhevnikova AN. *Appl Biochem Biotechnol.* 2003;109:285-301.

[49] Spyros A, Kimmich R, Briese BH, Jendrossek D. *Macromolecules*, 1997;30:8218-8225.
[50] Hocking PJ, Marchessault RH, Timmins MR, Lenz RW, Fuller RC, *Macromolecules*, 1996;29:2472-2478.
[51] Bonartseva GA, Myshkina VL, Nikolaeva DA, Rebrov AV, Gerasin VA, Makhina TK. *Mikrobiologiia* 2002;71:258-263. in Russian.
[52] Kramp B, Bernd HE, Schumacher WA, Blynow M, Schmidt W, Kunze C, Behrend D, Schmitz KP. *Laryngorhinootologie*, 2002;81:351-356. in German.
[53] Malm T, Bowald S, Karacagil S, Bylock A, Busch C. *Scand J Thorac Cardiovasc Surg.* 1992;26:9-14.
[54] Malm T, Bowald S, Bylock A, Saldeen T, Busch C. *Scand J Thorac Cardiovasc Surg.* 1992, 26:15-21.
[55] Malm T, Bowald S, Bylock A, Busch C. *J Thorac Cardiovasc Surg.*, 1992;104:600-607.
[56] Malm T, Bowald S, Bylock A, Busch C, Saldeen T. *Eur. Surg. Res.*, 1994;26:298-308.
[57] Duvernoy O, Malm T, Ramström J, Bowald S. *Thorac Cardiovasc Surg.* 1995;43:271-274.
[58] Kawaguchi T, Tsugane A, Higashide K, Endoh H, Hasegawa T, Kanno H, Seki T, Juni K, Fukushima S, Nakano M. *J. Pharma. Sci.*, 1992;87:508-512.
[59] Baptist, JN. US Patent No. 3 225 766, 1965.
[60] Holmes P. *Biologically produced (R)-3-hydroxy-alkanoate polymers and copolymers.* In: Bassett DC (Ed.) *Developments in crystalline polymers.* Elsevier, London, 1988;2:1-65.
[61] Saad B, Ciardelli G, Matter S, Welti M, Uhlschmid GK, Neuenschwander P, Suterl UW. *J. Biomed. Mat. Res.*, 1996;30:429-439.
[62] Fedorov M, Vikhoreva G, Kildeeva N, Maslikova A, Bonartseva G, Galbraikh L. *Chimicheskie volokna* 2005;6:22-28. in Russian.
[63] Rebrov AV, Dubinskii VA, Nekrasov YP, Bonartseva GA, Shtamm M, Antipov EM. *Vysokomol. Soedin* 2002;44:347-351. in Russian.
[64] Kil'deeva NR, Vikhoreva GA, Gal'braikh LS, Mironov AV, Bonartseva GA, Perminov PA, Romashova AN. *Prikl. Biokhim. Mikrobiol.* 2006;42:716-720. in Russian.
[65] Solheim E, Sudmann B, Bang G, Sudmann E. *J. Biomed. Mater. Res.* 2000;49:257-263.
[66] Bostman O, Pihlajamaki H. *Biomaterials*, 2000;21:2615-2621.

[67] Lickorish D, Chan J, Song J, Davies JE. *Eur. Cell. Mater.*, 2004;8:12-19.
[68] Khouw IM, van Wachem PB, de Leij LF, van Luyn MJ. *J. Biomed. Mater. Res.*, 1998;41:202-210.
[69] Su SH, Nguyen KT, Satasiya P, Greilich PE, Tang L, Eberhart RC. *J. Biomater. Sci. Polym. Ed.* 2005;16:353-370.
[70] Lobler M, Sass M, Schmitz KP, Hopt UT. *J. Biomed. Mater. Res.*, 2003;61:165-167.
[71] Novikov LN, Novikova LN, Mosahebi A, Wiberg M, Terenghi G, Kellerth JO. *Biomaterials*, 2002;23:3369-3376.
[72] Cao W, Wang A, Jing D, Gong Y, Zhao N, Zhang X. *J. Biomater. Sci. Polymer Ed.*, 2005;16:1379-1394.
[73] Wang YW, Yang F, Wu Q, Cheng YC, Yu PH, Chen J, Chen GQ. *Biomaterials* 2005;26.755-761.
[74] Ostwald J, Dommerich S, Nischan C, Kramp B. *Laryngorhinootologie*, 2003;82:693-699 in German.
[75] Wollenweber M, Domaschke H, Hanke T, Boxberger S, Schmack G, Gliesche K, Scharnweber D, Worch H. *Tissue Eng.* 2006;12:345-359.
[76] Wang YW, Wu Q, Chen GQ. *Biomaterials*, 2004;25:669-675.
[77] Nebe B, Forster C, Pommerenke H, Fulda G, Behrend D, Bernewski U, Schmitz KP, Rychly J. *Biomaterials* 2001;22:2425-2434.
[78] Qu XH, Wu Q, Chen GQ. *J. Biomater. Sci. Polymer Ed.*, 2006;17:1107-1121.
[79] Pompe T, Keller K, Mothes G, Nitschke M, Teese M, Zimmermann R, Werner C. *Biomaterials,* 2007;28:28-37.
[80] Deng Y, Lin XS, Zheng Z, Deng JG, Chen JC, Ma H, Chen GQ. *Biomaterials*, 2003;24:4273-4281.
[81] Zheng Z, Bei FF, Tian HL, Chen GQ. *Biomaterials*, 2005;26:3537-3548.
[82] Qu XH, Wu Q, Liang J, Zou B, Chen GQ. *Biomaterials.* 2006;27:2944-2950.
[83] Shishatskaya EI, Volova TG. *J. Mater. Sci-Mater. M.*, 2004;15:915-923.
[84] Fischer D, Li Y, Ahlemeyer B, Kriglstein J, Kissel T. *Biomaterials* 2003;24:1121-1131.
[85] Nitschke M, Schmack G, Janke A, Simon F, Pleul D, Werner C. *J. Biomed. Mater. Res.*, 2002;59:632-638.
[86] Chanvel-Lesrat DJ, Pellen-Mussi P, Auroy P, Bonnaure-Mallet M. *Biomaterials* 1999;20:291-299.

[87] Boyan BD, Hummert TW, Dean DD, Schwartz Z. *Biomaterials* 1996;17:137-146.
[88] Bowers KT, Keller JC, Randolph BA, Wick DG, Michaels CM. *Int. J. Oral. Max. Impl.*, 1992;7:302-310.
[89] Cochran D, Simpson J, Weber H, Buser D. *Int. J. Oral. Max. Impl.*, 1994;9:289-297.
[90] Sevastianov VI, Perova NV, Shishatskaya EI, Kalacheva GS, Volova TG. *J. Biomater. Sci. Polym. Ed.*, 2003;14:1029-1042.
[91] Seebach D, Brunner A, Burger HM, Schneider J, Reusch RN. *Eur. J. Biochem.*, 1994;224:317-328.
[92] Reusch RN. *Proc. Soc. Exp. Biol. Med.*, 1989;191:377-381.
[93] Reusch RN. *FEMS Microbiol. Rev.*, 1992;103:119-130.
[94] Reusch RN. *Can. J. Microbiol.*, 1995;41:50-54.
[95] Müller HM, Seebach D. *Angew Chemie*, 1994;32:477-502.
[96] Huang R, Reusch RN. *J. Biol. Chem.* 1996;271:22196-22201.
[97] Reusch RN, Bryant EM, Henry DN. *Acta Diabetol.*, 2003;40:91-94.
[98] Reusch RN, Sparrow AW, Gardiner J. *Biochim. Biophys. Acta*, 1992;1123:33-40.
[99] Reusch RN, Huang R, Kosk-Kosicka D. *FEBS Lett.*, 1997;412:592-596.
[100] Pavlov E, Zakharian E, Bladen C, Diao CTM, Grimbly C, Reusch RN, French RJ. *Biophysical J.*, 2005;88:2614-2625.
[101] Theodorou MC, Panagiotidis CA, Panagiotidis CH, Pantazaki AA, Kyriakidis DA. *Biochim. Biophys. Acta.*, 2006;1760:896-906.
[102] Wiggam MI, O'Kane MJ, Harper R, Atkinson AB, Hadden DR, Trimble ER, Bell PM. *Diabetes Care*, 1997;20:1347-1352.
[103] Larsen T, Nielsen NI. *J. Dairy Sci.*, 2005;88:2004-2009.
[104] Agrawal CM, Athanasiou KA. *J. Biomed. Mater. Res.*, 1997;38:105-114.
[105] Ignatius AA, Claes LE. *In vitro biocompatibility of bioresorbable polymers: poly(l, dl-lactide) and poly(l-lactide-co-glycolide).* 1996;17:831-839.
[106] Rihova B. *Adv Drug. Delivery Rev.*, 1996;21:157-176.
[107] Ceonzo K, Gaynor A, Shaffer L, Kojima K, Vacanti CA, Stahl GL. *Tissue Eng.* 2006;12:301-308.
[108] Chasin M, Langer R. (Eds), *Biodegradable Polymers as Drug Delivery Systems,* New York, Marcel Dekker, 1990.
[109] Johnson OL, Tracy MA. *Peptide and protein drug delivery.* In: Mathiowitz E, Ed. Encyclopedia of Controlled Drug Delivery. Vol 2. Hoboken, NJ: John Wiley and Sons; 1999, 816-832.

[110] Jain RA. *Biomaterials.* 2000;21:2475-2490.
[111] Gursel I, Hasirci V. *J. Microencapsul.*, 1995;12:185-193.
[112] Li J, Li X, Ni X, Wang X, Li H, Leong KW. *Biomaterials* 2006;27:4132-4140.
[113] Akhtar S, Pouton CW, Notarianni LJ. *Polymer*, 1992;33:117-126.
[114] Akhtar S, Pouton CW, Notarianni LJ. *J. Controlled Release*, 1991;17:225-234.
[115] Korsatko W, Wabnegg B, Tillian HM, Braunegg G, Lafferty RM. *Pharm. Ind.* 1983;45:1004-1007.
[116] Korsatko W, Wabnegg B, Tillian HM, Egger G, Pfragner R, Walser V. *Pharm. Ind.*, 1984;46:952-954.
[117] Kassab AC, Piskin E, Bilgic S, Denkbas EB, Xu K. *J. Bioact. Compat. Polym.*, 1999;14:291-303.
[118] Kassab AC, Xu K, Denkbas EB, Dou Y, Zhao S, Piskin E. *J. Biomater. Sci. Polym. Ed.*, 1997;8:947-961.
[119] Sendil D, Gursel I, Wise DL, Hasirci V. *J. Control. Release* 1999;59:207-217.
[120] Gursel I, Yagmurlu F, Korkusuz F, Hasirci V. *J. Microencapsul.*, 2002;19:153-164.
[121] Turesin F, Gursel I, Hasirci V. *J. Biomater. Sci. Polym. Ed.*, 2001;12:195-207.
[122] Turesin F, Gumusyazici Z, Kok FM, Gursel I, Alaeddinoglu NG, Hasirci V. *Turk. J. Med. Sci.*, 2000;30:535-541.
[123] Gursel I, Korkusuz F, Turesin F, Alaeddinoglu NG, Hasirci V. *Biomaterials*, 2001;22:73-80.
[124] Korkusuz F, Korkusuz P, Eksioglu F, Gursel I, Hasirci V. *J. Biomed. Mater. Res.*, 2001;55:217-228.
[125] Yagmurlu MF, Korkusuz F, Gursel I, Korkusuz P, Ors U, Hasirci V. *J. Biomed. Mater. Res.*, 1999;46:494-503.
[126] Khang G, Kim SW, Cho JC, Rhee JM, Yoon SC, Lee HB. *Biomed. Mater. Eng.*, 2001;11:89-103.
[127] Bonartsev AP, Bonartseva GA, Makhina TK, Mashkina VL, Luchinina ES, Livshits VA, Boskhomdzhiev AP, Markin VS, Iordanskii A.L. *Prikl. Biokhim. Mikrobiol.* 2006;42:710-715.
[128] Salman MA, Sahin A, Onur MA, Oge K, Kassab A, Aypar U. *Acta Anaesthesiol. Scand.*, 2003;47:1006-1012.
[129] Bonartsev AP, Livshits VA, Makhina TA, Myshkina VL, Bonartseva GA, Iordanskii AL. *Express Polymer Letters*, 2007;1:797-803.

[130] 130.Bonartsev AP, Postnikov AB, Myshkina VL, Artemieva MM, Medvedeva NA. *Am. J. Hypertension*, 2005;18:45-56.
[131] Bonartsev AP, Postnikov AB, Mahina TK, Myshkina VL, Voinova VV, Boskhomdzhiev AP, Livshits VA, Bonartseva GA, Iordanskii AL. *J. Clinical Hypertension*, 2007;9:A152-155.
[132] Pouton CW, Akhtar S. *Adv. Drug Deliver. Rev.*, 1996;18:133-162.
[133] Kots AY, Grafov MA, Khropov YV, Betin VL, Belushkina NN, Busygina OG, Yazykova MY, Ovchinnikov IV, Kulikov AS, Makhova NN, Medvedeva NA, Bulargina TV, Severina IS. *Br. J. Pharmacol.* 2000;129:1163-1177.

In: Biodegradable Polymers ...
Editors: A. Jimenez and G. E. Zaikov
ISBN 978-1-61209-520-2
© 2011 Nova Science Publishers, Inc.

Chapter 2

USE OF HYDROXYTYROSOL AS POLYPROPYLENE STABILIZER AND AS A POTENTIAL ACTIVE ANTIOXIDANT

M. Peltzer[1], N. López[2], L. Matisová-Rychlá[3], J. Rychlý[3] and A. Jiménez[1]

[1]University of Alicante, Analytical Chemistry, Nutrition and Food Sciences Department, P.O. Box 99, 03080 Alicante, Spain; mercedes.peltzer@ua.es
[2]AIMPLAS. Instituto Tecnológico del Plástico. C/ Gustave Eiffel, 4, 46980, Paterna, Valencia, Spain
[3]Polymer Institute, Slovak Academy of Sciences, Dubravska cesta 9842 36 Bratislava, Slovak Republic

ABSTRACT

The use of natural antioxidants as stabilizers in polymer formulations has raised some interest in recent years making attractive their use in packaging applications. One of the main sources of such phenolic compounds is olive fruit which contains a wide variety of bioactive components. Among them, hydroxytyrosol (HT) stands out as a high value compound due to its antioxidant character and beneficial properties to food and human health. The aim of this study is to characterize and evaluate the stabilizing performance of HT in polypropylene (PP) films as a potential polymer stabilizer and active additive. Oxidation Induction Time (OIT) of selected samples of polypropylene (PP) additivated with HT was determined by Differential Scanning Calorimetry (DSC) and Chemiluminescence (CL). In addition, the Onset Oxidation Temperature (OOT) and the apparent activation energy (E_a) were determined by DSC and thermogravimetric analysis (TGA), respectively. HT showed good performance as PP stabilizer at concentrations higher than 0.05 wt%, since it was observed an increment in the OIT, OOT and E_a values. HT at concentrations higher than 0.5 wt% could be considered as active additive in PP films.

Keywords: Hydroxytyrosol; natural antioxidants; polypropylene; oxigen induction time (OIT), chemiluminescence

INTRODUCTION

The use of natural antioxidants as stabilizers in polymer formulations has made them attractive in packaging applications [1-3]. These antioxidants are non-toxic compounds obtained from renewable resources, either from plants, herbs and spices such as carnosic acid and carvacrol [2,3], or directly from vegetable oils, such as tocopherols and hydroxytyrosol (3,4-dihydroxyphenylethanol) (HT) [4].

The olive fruit contains a wide variety of bioactive components. Among them, HT stands out as a high value compound due to its antioxidant character and beneficial properties [5,6]. HT is extracted from the solid by-product obtained during olive processing. It was found to play a role in enhancing the oxidative stability of olive oil with simultaneous positive effects on human health [4]. Therefore, it is a promising candidate to be used in active packaging systems, where HT could play a role on the material stabilization as well as on foodstuff protection. In some studies HT was already suggested as a potential natural additive for polymers and it was compared with α-tocopherol, a well-known natural antioxidant already used in polymer stabilization [7-9].

Thermal analysis has been extensively used for characterization of polymer materials and also to evaluate the stabilization performance of antioxidants and/or stabilizers. Differential Scanning Calorimetry (DSC) and Thermogravimetry (TGA) among others are widely employed in both scientific and industrial sectors. Some parameters such as apparent activation energies (E_a), oxidation onset temperatures (OOT) and oxidation induction time (OIT) can be calculated by using these techniques in order to evaluate the stabilization performance of additives [10].

Natural additives are also used in active packaging systems based on the incorporation of substances, in particular antimicrobials and antioxidant agents to the polymer matrix [11]. The incorporation of active additives leads to an increase in the shelf-life of foodstuff, and it will also allow to reduce the amount of the food additives. The aim of this study is to characterize and to evaluate the stabilizing performance of HT in polypropylene (PP) films as a potential polymer stabilizer and active additive.

EXPERIMENTAL

Compounding preparation

The materials used in this work were medium melt-flow PP without any additive in powder form. HT (50% purity) was supplied by Genosa S.A. (Málaga,Spain) and it was added to PP at 5 wt% at a first approach. Both components were then mixed in a turbomixer at 1000 rpm during 5 minutes. HT was disolved in ethanol, since it was too viscous to be directly added to the polymer. Powder PP, without any additive, was further added to the mixture in order to obtain different concentrations of HT in the final material, i.e. 0.01; 0.05; 0.1 and 0.5 wt%. Each mixture was processed in a twin-screw extruder (Coperion, Barcelona, Spain) with 25 mm diameter and screw lenght of 40 mm with several degasification zones. The screw configuration was optimized taking into account the specific mechanical energy needed to obtain a suitable dispersion of the additive into the PP matrix. Extrusion was performed at six temeperature zones, 190/200/195/195/195/190 °C at 200 rpm. The material was pelletized before preparing films.

Films preparation

Pellets were transformed into films by using a cast sheet extrusion line (Dr. Collin) with L/D ratio 25. This equipment is composed by a single screw extruder, a cast sheet die and a calender to cool and stretch the material. The temperature profile in the extruder was 165/175/185/185 °C and the die temperature was 185 °C. These parameters were selected after optimization to control films dimension, crystallization and orientation. The extrusion rate was 40-50 rpm and the calender temperature was 40-50 °C. The film thickness, 78.4 ± 6.2 µm, was measured with a Digimatic Micrometer Series 293 MDC-Lite (Mitutoyo, Japan).

DSC studies in oxydant atmosphere: determination of OIT and OOT

Samples weighing approximately 3 mg were put into aluminum pans for DSC testing in a TA Instruments Q2000 (New Castle, DE, USA). OIT tests were performed by heating the sample at 100 °C min^{-1} in inert atmosphere (N$_2$, 50 mL min^{-1}) from room temperature up to 200 °C. After 5 minutes at this temperature the atmosphere was changed to oxygen (50 mL min^{-1}) and the isothermal heat evolution was monitored until the detection of the exothermic peak indicating the oxidation reaction. In order to determine OOT, samples were heated in oxygen atmosphere (50 mL min^{-1}) at 10 °C min^{-1} from 30 °C to the temperature where the exothermic oxidation peak was observed. OIT and OOT were calculated as the temperature for the intersection between the baseline and the slope of the exothermic peak in each case.

Chemiluminescence studies

Chemiluminescence experiments were performed on a photon-counting instrument Lumipol 2 manufactured by the Polymer Institute of Slovak Academy of Sciences, Bratislava, Slovakia. Tests were performed with oxygen flow above the sample surface at 3 L h^{-1}. Circle samples (9 mm diameter) were prepared from PP films and they were placed into aluminum pans. The instrument dark count rate was 2-4 counts s^{-1} at 40 °C. Isothermal tests in pure oxygen were performed at 150 °C. Induction time was then calculated as the intersection between the tangent of the signal and the time axes.

TGA studies in inert and oxidant atmospheres

TGA tests were carried out by using a Mettler Toledo TGA/SDTA 851 instrument. Dynamic tests were run at 10° C min^{-1} from 30 to 550°C in inert (N$_2$) or oxidant atmosphere (O$_2$) with flow rate in both cases 50 mL min^{-1}. Samples were accurately weighed around 5 mg, in a standard aluminum pan without any previous treatment.

RESULTS

DSC studies: determination of OIT (min) and OOT (°C)

Figure 1 shows the dynamic DSC curves in oxygen for all materials in the calculation of OOT. It can be observed that the melting peak of PP at T$_m$ = 163 °C did not change with the addition of HT but a clear shift to higher temperatures in samples stabilized with HT at concentrations higher than 0.05wt% was observed.

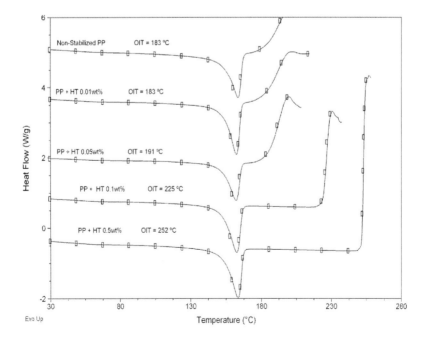

Figure 1. Determination of OOT (°C) for PP stabilized with HT.

Table 1 shows the results obtained for OIT at 150 °C and 200 °C for PP stabilized with different amounts of HT. A clear increase in this parameter with the concentration of HT was observed and once again, initial concentrations of antioxidant higher than 0.05 wt% showed a considerable increase in the OIT value. Both temperatures were studied to evaluate the effectiveness of the HT at temperatures either well below the polymer melting point or in the melting state.

As expected at low temperatures the induction times became too long to be measured by a conventional DSC instrument since it is not practical. In addition testing of oxidations parameters has been somewhat discussed since the heat flow is too slow and this reduces the method sensitivity. Therefore, the evaluation of the antioxidant performance of HT at low temperatures should be studied by using techniques with good sensitivity under these conditions. CL, which appears to become a popular technique for testing polymers stability, allow us the determination of induction and maximum oxidation times at temperatures below the polymer melting point.

Table 1. Oxygen induction time values for PP stabilized with HT

Sample	OIT (min) at 150 °C	OIT (min) at 200 °C
Non-stabilized PP	6.2	0.4
PP+HT 0.01wt%	7.5	0.4
PP+HT 0.05wt%	37.2	0.9
PP+HT 0.1wt%	190	7.2
PP+HT 0.5wt%	>200	37.4

Chemiluminescence studies at 150 °C

Figure 2 shows results obtained for the CL emission for all samples in oxygen at 150 °C. Chemiluminescence in polymers arises by the self-recombination of secondary peroxyl radicals according to the Russel mechanism (Scheme 1):

$$2\ RO_2^\bullet \longrightarrow ROH + cetona^{\bullet *} + O_2^{\bullet *} \qquad \text{Scheme 1}$$

The presence of an inhibitor (antioxidant) scavenges those peroxyl radicals until its total consumption. Therfore, the typical auto-accelerating chemiluminescnce vs time curves are obtained, where the light intensity is low when inhibitor is present and increases suddenly when it is gradually consumed by reaction with peroxyl radicals. Therefore, the antioxidant efficiency can be related to the time needed for the CL signal to increase [12]. In this case, this induction time was calculated by the intersection between the baseline and the tangent of the increase in CL intensity (Figure 2). A clear shift in the beginning of CL intensity to higher times when the concentration of HT increases was observed. As it was obtained in DSC tests, PP stabilized with low concentrations of HT did not show significant differences with results obtained with the pure polymer. But, at HT concentrations higher than 0.05 wt% clear differences were observed. Results also showed a decrease in CL intensity with the increasing amount of antioxidant. This effect could be explained by the formation of several carbonyl compounds during the oxidative degradation of polypropylene, such as ketones, esters, lactones and carboxilic acids [13]. It has been reported that the CL intensity is proportional to the accumulation of carbonyl species formed during oxidation [14].

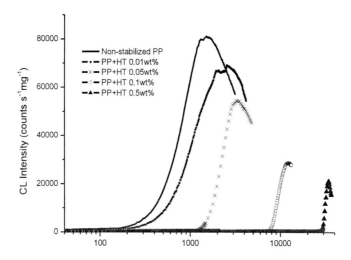

Figure 2. CL curves for PP stabilized with HT.

Isothermal CL-curves could be fitted with *Equation 1* as it was indicated elsewhere [15].

$$I = \frac{A\exp(-k_1)}{[1 + Y\exp(-k_2 t)]^2} \quad (1)$$

The main advantage of this equation lies on the direct relation with the oxidation process. Thus, k_1 corresponds to the rate constant of isolated hydroperoxides decomposition while k_2 is related to the rate constant of associated hydroperoxides decomposition. The induction time can be also determined by the parameter Y, which is the ratio of the maximum $[POOH]_\infty$ (at infinity time) and the initial concentration of hydroperoxides $[POOH]_0$ (*Equation 2*).

Eq. 2 $$Y = \left\{ \frac{[POOH]_\infty}{[POOH]_0} \right\} - 1$$

The initial concentration of hydroperoxides should be considered as an indication of the polymer quality, including all structural defects in the polymer due to processing, reactions of catalyst residuals, etc. These structural defects are points where degradation processes start by their conversion to hydroperoxides. The higher the initial concentration of these structural defects the lower the value of Y. Table 2 shows the value of the different parameters obtained from CL curves.

Table 2. Induction time (t_{ind}), Maximum intensity (I_{max}) and parameters obtained from *Eq.1*

Material	t_{ind} (min)	I_{max} (x10^{-4})	A	Y	k_1 (x10^4)	k_2 (x10^3)
Non-stabilized PP	5.2 ± 0.1	8.1	1.2x10^5	11	2.6	3.8
PP+HT 0.01wt%	8.1 ± 1.8	6.9	9.1x10^4	6,9	1.1	2.4
PP+HT 0.05wt%	25 ± 1	5.5	1.0x10^5	118	1.8	2.4
PP+HT 0.1wt%	134 ± 1	2.9	3.6x10^4	4.7x10^4	2.0	1.2
PP+HT 0.5wt%	504 ± 21	2.1	1.8x10^7	8.3x10^{10}	1.9	0.8

Hydroperoxydes could be considered as intermediates in polyolefins oxidation [16]. As expected, the value of Y was higher with the increasing HT initial concentration, meaning that the antioxidant protected the polymer during processing since there was a decrease in the number of initial hydroperoxides. It could be also observed that the rate constant of associated hydroperoxides decomposition (k_2) decreased for stabilized samples resulting in lower degradation since the presence of associated hydroperoxydes leads to a faster polymer oxidation.

The increase in induction time determined by DSC and CL represents that there is certain amount of antioxidant remained in the polymer matrix. Results obtained in the different themo-oxidative CL tests showed that HT in concentrations higher than 0.05wt% could be used as PP stabilizer since significant increases were observed in both parameters, induction time and temperature. But this study should be completed by the calculation of another parameter very useful to evaluate stability of polymers, such as apparent activaton energies (E_a) of the degradation process [10].

TGA studies in inert and oxydant atmosphere

Figure 3 shows TG curves for three of the studied materials in inert atmosphere. It can be observed that their degradation is a one step process, typical from polyolefins. The mass loss of additives in these cases could not be observed because of their low concentrations. From these curves, the initial and maximum rate degradation temperatures were determined and are indicated in Table 3. A sligth increase in the initial degradation temperature ($T_{5\%}$) with the increasing amount of HT was observed, while no significant variations in the maximum degradation rate temperature (T_{max}) values from the different samples were observed.

Table 3. Degradation temperatures, $T_{5\%}$ and T_{max} (°C) for PP with different HT concentrations

Sample	T_{ini} (°C)	T_{max} (°C)
Non-stabilized PP	436	463
PP + HT 0.01wt%	436	461
PP + HT 0.05wt%	437	463
PP + HT 0.1wt%	443	464
PP + HT 0.5wt%	442	463

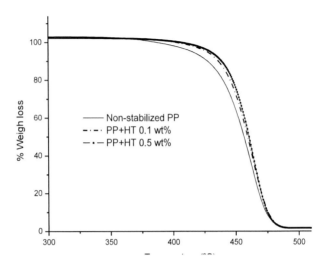

Figure 3. TGA curves for PP stabilized with HT in inert atmosphere.

Figure 4 shows TGA and DTG curves for all the materials studied in oxidant atmosphere. It can be observed that curves shifted to higher temperatures in those samples with higher amount of antioxidants.

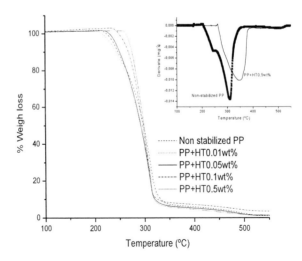

Figure 4. - TGA and DTG curves for PP stabilized with HT in oxygen atmosphere (50 mL min^{-1}).

Table 4 shows temperatures for the initial and maximum degradation rate in oxidant atmosphere for all materials.

Table 4. Initial degradation temperature $T_{5\%}$ (°C) and maximum degradation rate temperature T_{max} (°C) in oxygen

Sample	$T_{5\%}$ (°C)	T_{max} (°C)
Non-stabilized PP	174	314
PP + HT 0.01wt%	176	324
PP + HT 0.05wt%	194	331
PP + HT 0.1wt%	224	323
PP + HT 0.5wt%	257	323

It has been stated that thermal analysis methods are actually used for evaluating the thermal stability of polymers and the efficiency of antioxidants in the case of studying degradation in oxidant atmosphere [17]. Kinetic parameters were calculated, in particular E_a, at different heating rates (β) = 5, 10, 15 and 20 °C min^{-1} under oxygen atmosphere and using the Flynn-Wall method [18]. This method gives an easy estimation of E_a and the obtained values are shown in Table 5.

Table 5. Apparent activation energy (E_a, kJ mol^{-1}) calculated by Flynn-Wall method

Sample	Non-stabilized PP	PP + HT 0.1wt%	PP + HT 0.5wt%
E_a (kJ mol^{-1})	49.5 ± 3.3	71.7 ± 2.8	75.4 ± 8.2

It can be observed that the apparent activation energy increases for stabilized samples from these results it was demonstrated that HT could be used to stabilize PP during processing by the clear increase in E_a values. The role of HT as hydrogen peroxide scavenger was already stated [19], and it could lead to the oxidative degradation rate. This antioxidant performance was increased by the formation of dimers and quinone products during the autooxidation of HT [19,20], which were said to be also good antioxidants by themselves.

CONCLUSION

HT could be considered as a good alternative to synthetic antioxidants commonly used in PP formulations, since some stabilization in the polymer matrix was observed after the addition of relatively low amounts of this compound. Initial concentrations higher than 0.05 wt% were observed to be enough to stabilize PP. In addition, concentrations higher than 0.5 wt% could be considered to be used as active additive in PP films, since the increase in oxidation induction parameters, such as OIT, indicates that residual amounts of stabilizer in were remaining in materials after processing. Further studies will be carried out in order to study the release of HT to food samples and to evaluate its antioxidant performance.

REFERENCES

[1] Standberg C, Albertsson AC. *J. Appl. Polym. Sci.*, 2005;98:2427-2439.
[2] Bentayeb K, Rubio C, Battle R, Nerin C. *Anal. Bioanal. Chem.*, 2007;389:1989-1996.
[3] Singh G, Kiran S, Marimuthu P, Isidorov V. *J. Agric. Food Sci.*, 2008;88:280-289.
[4] Fernández-Bolaños J, Rodríguez G, Rodríguez R, Heredia A, Guillén R, Jiménez A. *J. Agric. Food. Chem.* 2002;50:6804-6811.
[5] Sousa A, Ferreira ICFR, Barros L, Bento A, Pereira JA. *LWT-Food Sci. Technol.*, 2008;41:739-745.
[6] De Jong S, Lanari MC. *Food Chem*, 2009;116:892-897.
[7] Al-Malaika S, Issenhuth S. *J. Vinyl Addit. Technol.*, 2004;10:52-56.
[8] Peltzer, M., Jiménez, A. *J. Thermal Anal. Calorim.*, 2009;96:243-248.
[9] Peltzer M, Wagner J, Jiménez A. *J. Thermal Anal. Calorim.*, 2007;87:493-497.
[10] Dobkowski Z. *Polym. Degrad. Stabil.*, 2006;91:488-493.
[11] Mascheroni E, Guillard V, Nalin F, Mora L, Piergiovanni L. *Food Eng.*, 2010;98(3):294-301.
[12] Cerruti P, Malinconio M, Rychly J, Matisova-Rychlá L, Carfagna C. *Polym Degrad Stabil*, 2009;94:2095-2100.
[13] Philippart JL, Sinturel C, Arnaud R, Gardette JL. *Polym Degrad Stabil*, 1999;64:213-225.
[14] Blakey I, George GA. *Macromolecules*, 2001;34:1873-1880.

[15] Matisová-Rychlá L, Rychlý J. *J. Polym. Sci. Part A: Polym. Chem.*, 2004;42:648-660.
[16] Gugumus F. *Polym. Degrad. Stabil.* 1995;49(1):29-50.
[17] Zhu L, Chen J, Xu L, Lian X, Chen M. *Polym Degrad Stabil*, 2009;94:906-913.
[18] Flynn JH, Wall LA. *J. Polym. Sci. Part B: Polym. Lett.* 1966;4:323-328.
[19] De Lucia M, Panzella L, Pezzella A, Napolitano A, D'Ischia AM. *Tetrahedron*, 2006;62:1273–1278.
[20] Vogna D, Pezzella A, Panzalla L, Napolitano A, D'Ischia M. *Tetrahedron Letters*, 2003;44-45:8289-8292.

In: Biodegradable Polymers ...
Editors: A. Jimenez and G. E. Zaikov

ISBN 978-1-61209-520-2
© 2011 Nova Science Publishers, Inc.

Chapter 3

MECHANICAL PROPERTIES OF DIMER FATTY ACID-BASED POLYAMIDES BIOCOMPOSITES

Elodie Hablot[1], Rodrigue Matadi[2], Said Ahzi[2] and Luc Avérous[1],*

[1]Laboratoire d'Ingénierie des Polymères pour les Hautes Technologies, ECPM-LIPHT, Université de Strasbourg, 25 rue Becquerel, 67087 Strasbourg, Cedex 2, France

[2]Institut de Mécanique des Fluides et des Solides; IMFS Université de Strasbourg, Strasbourg, 2 Rue Boussingault; 67000 Strasbourg, France

ABSTRACT

Nowadays, replacing petroleum-based materials with renewable resources is a major concern in terms of both economical and environmental points of view. In good agreement with this emergent concept of sustainable development, this work deals with the study of innovative green polyamides (DAPA) based on rapeseed oil-dimer fatty acid reinforced with cellulose fibres (CF). Mechanical properties of DAPA and its biocomposites were examined. Tensile tests were used to follow the effect of strain rate, temperature and filler content on the

---

* Tel.:+ 33 368852784 - Fax: + 33 368852716 - AverousL@unistra.fr

Young modulus and the yield stress. Tensile tests revealed that the Young modulus and the yield stress highly increased with increasing CF concentration. Halpin-Tsai and Eyring's micro-mechanical models were found to successfully predict the Young modulus and the yield stress respectively, of DAPA and DAPA/Cellulose biocomposites. These results clearly highlighted the attractive properties of DAPA-based biocomposites and demonstrated that these materials are good alternatives to conventional composites.

Keywords: Short-fibre composites, Bio-polyamides, Cellulose fibres, Mechanical properties, Modelling

INTRODUCTION

The use of renewable materials to produce existing chemicals or to create original products is the subject of a large number of academic and industrial research projects. As an example, vegetable oils are expected to be an ideal alternative chemical feedstock, since they are produced in abundance throughout the world. Moreover, they show the particularity to contain lots of active chemical units that can be used for polymerization as bio-based monomers or building blocks [1-6].

Bio-based polyamides (excluding proteins and polypeptides) remain little-developed. One easy route to produce bio-based polyamides is by using dimers of fatty acids, as shown in Figure 1.

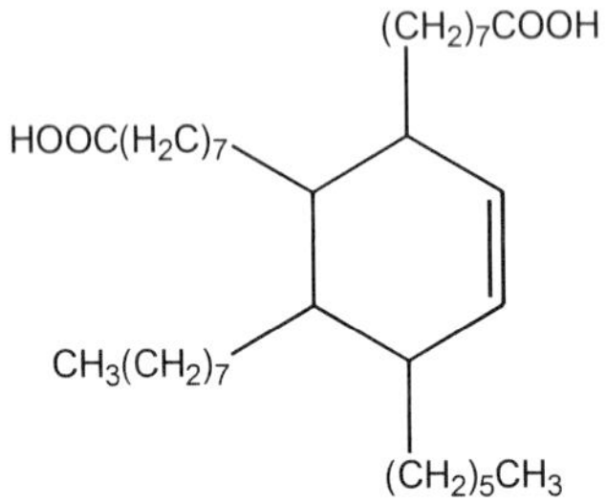

Figure 1. Fatty dimer acid structure.

They are well-known and commercially available products and they are good candidates to synthesize bio-based thermoplastic polyamides [7]. Dimer fatty acid-based polyamides (DAPA), due to their special molecular structure based on the dimer acid, show relatively good performance for such purpose. But their properties are still far from those of conventional polyamide thermoplastics [8]. Then, an effective approach to improve the DAPA behaviour is the incorporation of fibres into the matrix. In order to successfully meet the environment and recycling problems, agro-based fibres are currently being introduced in different industrial sectors (e.g. automotive) to replace e.g., common glass-fibers. Because of their relative low cost, low density and ability to recycle [9-10], cellulosic fibers seem to be good candidates to reinforce DAPA and to prepare biocomposites essentially based on renewable materials, on agreement with the sustainable development concept. In previous works [11-12], we have focused our research on the processing, characterisation and structure of DAPA/Cellulose (DAPAC) composites. We propose herein to extent this previous work to the study of the mechanical properties of DAPAC. Mechanical models were also developed to predict the Young modulus and the yield stress of DAPA and DAPAC.

EXPERIMENTAL

Materials

Dimer acid used in this synthesis was supplied by Prolea (France) as a yellowish, viscous liquid at room temperature with the dimer, trimer and monomer content at 96.6, 2.8 and 0.6 %wt, respectively. 1,6-diaminohexane was purchased from Sigma Aldrich. All chemicals were used as received.

Pure cellulose fibers (CF) from leafwood (Arbocel B400) were supplied by JRS (Rosenberg, Germany). They showed an average length and thickness of 900 and 20 µm, respectively. In a previous paper, Amash et al [13] reported some characteristics of these fibers. According to authors, the density of fibers was 1.50 and the cellulose content was higher than 99.5%.

Polyamide Synthesis

In agreement with previous publications [11-12], DAPA was synthesized by condensation polymerization. Dimer acid was charged in a 1L five-necked round-bottom flask equipped with a mechanical stirrer, a thermometer, a nitrogen inlet, a Dean-Stark apparatus and a dropping funnel. Dimer acid was heated to 343 K under stirring and nitrogen atmosphere. 1,6-diaminohexane equivalent to total acid was added by using the dropping funnel for 10 min. The mixture was further heated gradually up to 453 K within 3 hours. The completion of the polymerization was checked by amine and acid titration. The reaction was stopped after the consumption of more than 99% of the acid functionalities. The final acid-based thermoplastic polyamide was then discharged from the flask.

DAPA is yellowish, transparent and flexible at ambient temperature with a molecular weight of 14,000 g mol^{-1}, a glass transition temperature of 263 K and a melting temperature of 354 K.

Biocomposites Processing

Prior to blending, fibers and DAPA were dried in an air-circulating oven at 383 K for 1h and 343 K for 1h, respectively. DAPA and different amounts of CF were directly added to the feeding zone of an internal mixer (Counter-rotating mixer Rheocord 9000, Haake, USA) at 373 K, with a rotation speed of 50 rpm and a processing time not exceeding 8 min, in order to avoid DAPA degradation. Four biocomposites (DAPAC) were prepared with 2, 5, 10, 15 and 20 %wt of CF content to form DAPAC2, DAPAC5, DAPAC10, DAPAC15 and DAPAC20, respectively. After melt-processing, composites were compression-moulded, to obtain plates, with a hot press at 403 K by applying 200 MPa for 5 min. The moulded specimens were then quenched between two steel plates for 10 min. Optical micrographs of thin sheets of DAPAC20 showed very good fiber dispersion into the DAPA matrix (Figure 2).

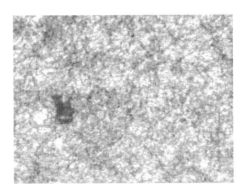

Figure 2. Optical micrograph (Magnification X2) of DAPAC20.

Characterization

Tensile tests were performed with an Instron tensile testing machine Model 4204 at different temperatures ranging from 298 to 333 K, and different strain rates ranging from 10-4s-1 to 1s-1. Test samples were held at those temperatures for 40 min to ensure equilibration of the samples for those tests conducted at elevated temperatures. The yield stress is defined in the traditional sense for polymers as the first maximum stress on the material right after the elastic point on the stress-strain curve. Therefore, even if tensile tests are carried out at temperatures around T_g and beyond, the real plastic deformation is controlled by tensile loading and unloading tests (Figure 3).

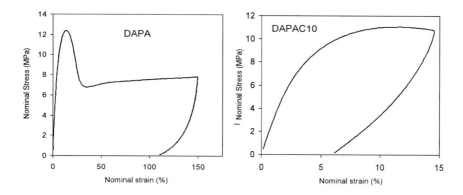

Figure 3. Loading and unloading curves of DAPA and DAPAC10 at strain rate of $10^{-2}s^{-1}$ and 298 K.

RESULTS AND DISCUSSION

Tensile Test Results

The stress-strain response, under tensile loading, of DAPA and DAPAC at 298 K is shown in Figure 4. The shape of the stress-strain curves of these materials looked like to the classical form of stress-strain curves of thermoplastic polymers. Moreover, a significant increase in the Young's modulus and yield stress of DAPAC was observed, since it increased from 181 MPa (without CF) to 707 MPa (with 20 %wt of CF) and from 11.4 (without CF) to 19.1 MPa (with 20 wt% of CF), respectively.

The Young's modulus and yield stress increased probably due to the good dispersion of the fillers and good interfacial interactions between fibers and the matrix. In parallel, the decreases in strain at break (from 600% without CF to 10% with 20 %wt of CF) were related to the stresses caused by the cellulose fiber. In fact, because of the modulus gap between cellulose and the matrix, stresses concentrations are located at the filler/matrix interfaces. Under tensile loading, crazes and cracks can be initiated at those weak points.

Young Modulus Modelling

Considering the DAPA/cellulose composites as two phase materials, the Young's modulus can be estimated by addition of the contribution of both phases. One of the most useful calculation methods is the classical bounds theory of Voigt [14] (upper bound) and Reuss [15] (lower bound), given by Equations 1 and 2, respectively.

$$E_c = f_D E_m + f_C E_f \qquad (1)$$

$$E_c = (f_D / E_m + f_C / E_f)^{-1} \qquad (2)$$

where E_m and E_f are respectively the elastic tensile modulus of DAPA matrix and cellulose fibers. f_D and f_c are respectively the volume fractions of DAPA matrix and the cellulose filler. The main mechanical properties are reported in Table 1.

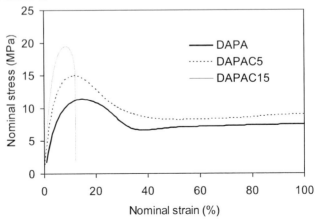

Figure 4. Stress-strain curves of pure DAPA and DAPAC with different filler contents.

Halpin and Tsai [16] developed a semi-empirical equation to approximate the Young's modulus of composite materials. The Halpin-Tsai formulation can be written according to Equation 3, where η is given by Equation 4.

$$\frac{E}{E_m} = \frac{1+\xi\eta\ f_C}{1-\eta\ f_C} \quad (3)$$

$$\eta = \frac{(E_f/E_m)-1}{(E_f/E_m)+\xi} \quad (4)$$

E represents the longitudinal modulus (E_L) or the transverse modulus (E_T). ξ is a measurement of reinforcement and depends on the fiber geometry. For the transverse modulus E_T satisfactory results have been obtained with $\xi = 2$. For the longitudinal modulus E_L, the fibers aspect ratio ξ was given by Equation 5 where L and D represent their length and diameter, respectively.

$$\xi = 2 \times \frac{L}{D} \quad (5)$$

Table 1. Materials mechanical properties

Materials	Density (g/cm^3)	Longitudinal modulus (GPa)	Transverse modulus (GPa)
Cellulose	1.50	80	6.70
DAPA	0.90	1.8×10^{-1}	1.8×10^{-1}

When the fibers are randomly dispersed in the matrix, the composite modulus is given by Equation 6 [17].

$$E_c = \frac{3}{8}E_L + \frac{5}{8}E_T \quad (6)$$

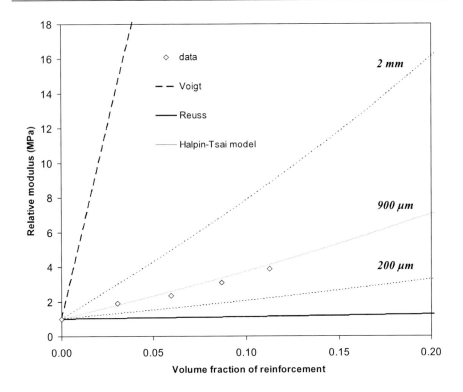

Figure 5. Experimental modulus values compared to theoretical predictions obtained with different models (Voigt, Reuss, and Halpin-Tsai).

Both classical bounds and Halpin-Tsai equations have been used to estimate the Young's modulus. The filler weight fraction was converted into volume fraction by using Equation 7.

$$f_C = w(w + (1-w)\rho_C / \rho_D) \qquad (7)$$

where w, ρ_C and ρ_D are the weight fraction, cellulose density and DAPA density respectively. Figure 5 represents the models predictions of the Young's modulus for DAPA/cellulose composites.

As expected the experimental data ranged from the upper and the lower bound models. The Voigt bound overestimates the Young's modulus, while Reuss bound underestimates it. But experimental results were closer to Reuss predictions, due to the large contrast between the Young's modulus of pure

DAPA and cellulose. Figure 5 also depictes the Halpin-Tsai model results for different fiber dimensions. It is observed that the Halpin-Tsai model provides a good fit of experimental data for $\xi = 90$, corresponding to the ξ value given by the fiber dimensions. These different models assumed that the interface is ideal without slip or chemical reactions, and that the fibers are well-dispersed in the matrix. But the studied systems did not show a perfect adhesion between the DAPA matrix and cellulose fibers. The best fit of experimental results suggested that the enhancement of the Young's modulus was more related to the fibers dispersion than to the strong interactions between the fibers and DAPA matrix. But this model could not be a perfect model for the full description of complex physics and chemistry controlling the behaviour of DAPA/cellulose composite. But the Halpin-Tsai model helped to obtain a good estimation of stiffness in the case of short fibre composites [18].

Yield Stress Modelling

It is well known that the yield stress of thermoplastic vanishes at temperatures higher than T_g [19-21]. In this study, the T_g of DAPA and DAPAC ranges between 285 and 289 K. According to the results reported in Figure 4, the stress-strain curves of both DAPA and DAPAC composites show a definite upper yield point followed by softening and strain hardening. As shown in previous works [22-23], even if the experimental temperature is higher than T_g, the behaviour of these materials into the rubbery region are not observed. Therefore, the cooperative model [24] below T_g has been used in this work. The temperature and strain rate dependence on yield behaviour of various polymers can be described by the following cooperative model [19-21] where the yield stress σ_y is expressed below T_g in Equation 8.

$$\frac{\sigma_y}{T} = \frac{\sigma_i(0) - m \cdot T}{T} + \frac{2k}{V_{eff}} \sinh^{-1}\left(\frac{\dot{\varepsilon}}{\dot{\varepsilon}_0 \exp\left(-\frac{\Delta H_{eff}}{kT}\right)}\right)^{1/n} \qquad (8)$$

Where T is the absolute temperature, k the Boltzmann constant, V_{eff} the effective activation volume, ΔH_{eff} the effective activation energy, ☐ the strain rate, ε_0 a pre-exponential constant, $\sigma_i(0)$ the internal stress at 0 K. Another important parameter is m defined as $\sigma_i(0)/T^*$, T^* being the compensation temperature while n describes the cooperative character of the yield process [24]. By following some recent works [25-26], an averaging approach (known as the Takayanagi model [27]) is used to estimate the effective activation volume and effective activation energy (Equation 9).

$$\Delta H_{eff} = \frac{\varphi \cdot \Delta H_C \cdot \Delta H_D}{\Omega \cdot \Delta H_D + (1-\Omega) \cdot \Delta H_C} + (1-\varphi) \cdot \Delta H_D$$

$$V_{eff} = \frac{\varphi \cdot V_C \cdot V_D}{\Omega \cdot V_D + (1-\Omega) \cdot V_C} + (1-\varphi) \cdot V_D$$

(9)

Where ΔH_D and V_D are the activation energy and activation volume of DAPA, respectively. ΔH_C and V_C are the activation energy and activation volume of cellulose fillers, respectively. φ and Ω are parameters related to the volume fractions of the DAPA matrix (f_D) and cellulose fibres (f_C) which are given by Equation 10.

$$\begin{cases} f_C = \varphi \cdot \Omega \\ f_D = 1 - \varphi \cdot \Omega \end{cases}$$

(10)

The volume fraction of cellulose is determined from the cellulose content in wt% (w), with Equation 7.

Equations 8 and 9 allow modeling the yield behaviour of DAPA and DAPAC. The biocomposite is considered a two phase material. The yield processes in each phase are described by the activation parameters of the matrix (ΔH_D and V_D) and those of cellulose fillers (ΔH_C and V_C).

Both Ree-Eyring and Richeton equations [24,28] are used to calculate the model parameters for DAPA. Both cooperative and the Ree-Eyring models consider the same β-activation energy. The activation energy and the activation volume of DAPA are determined with an asymptotic development of the Ree-Eyring hypoythesis given by Equation 11.

$$\sigma_y = \frac{2\Delta H}{V} + \frac{2kT}{V}\ln\left(\frac{2\dot{\varepsilon}}{\dot{\varepsilon}_0}\right) \quad (11)$$

Table 2. Cooperative model parameters

Parameters	Value
DAPA	
N	6
Φ	0.81
$\dot{\varepsilon}_0$	8.6E+22
$\sigma i(0)$ (MPa)	1.9
M	0.0067
$\rho_m (g/cm^3)$	0.90
$T_{ref}(K)$	313
$V_D (m^3)$	2.50E-27
ΔH_D (kJ/mol)	138.99
DAPAC	
$\rho_c (g/cm^3)$	1.5
$V_C (m^3)$	9.52E-28
ΔH_C (kJ/mol)	351.46

Note: The model parameters for DAPA and DAPAC are reported in Table 3.

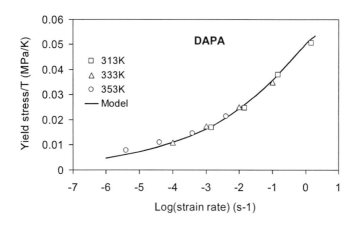

Figure 6. Master curve of DAPA at a reference temperature of 313K.

A plot of temperature-normalized yield stress (σ_y/T) vs. logarithmic strain rate sets a series of parallel, linear isotherms with a constant activation energy (ΔH) and activation volume (V), where R is the gas constant. From the slope and interception of such isotherms, ΔH and V could be computed. After the calculation of the activation parameters, a reference temperature ($T_{ref} = 313K$) was chosen. The horizontal and vertical shifts allowed to build a master curve at this temperature. The others parameters m, n, $\dot{\varepsilon}_0$, $\sigma_I(0)$ were obtained from the best fit of the master curve [25]. Figure 6 shows the master curve of DAPA built at 313 K.

By knowing the activated parameters of DAPA and the master curve of DAPAC at different cellulose concentrations, the effective activated parameters for DAPAC can be calculated from Eq. 8. From the ΔH_{eff} value, the parameters ΔH_D and ΔH_C could be estimated at a given cellulose content. The cooperative model parameters (V_{eff}, n, $\dot{\varepsilon}_0$) were then calculated from a fit of the master curve. Finally, the parameters V_D and V_C were derived from Equation 3. The model parameters of DAPA and DAPAC are shown in Table 2. The values reported mentioned a slight increase in the activation energy with the cellulose content, and a slight decrease of the activation volume with increasing cellulose content. These results are in agreement with the experimental results where the yield stress increased with the cellulose concentration.

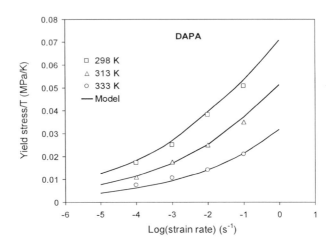

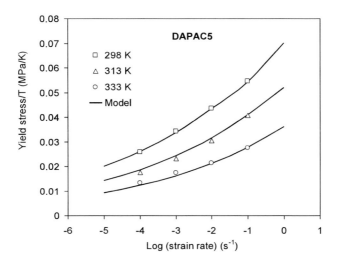

Figure 7. Yield stress/Temperature vs strain rate of DAPA.

Table 3. Variation of effective parameters with the cellulose content

Cellulose content (wt %)	Volume fraction (x10³)	ΔH_{eff} (kJ/mol)	V_{eff} (x10²⁷) (m³)
0	-	139	2.50
5	3.17	144	2.32
15	9.55	154	2.25

On a previous publication, we have shown that T_g increased with the addition of fibers [13]. This observation correlated to the increase in Yield Stress validated the fact that the addition of cellulose fibers leads to a reduction in the DAPA mobility in agreement with the behaviour of most of the cellulose-based composites [9]. Figure 7 shows the reduced yield stress (σ_y/T, versus log (strain rate)) plots for DAPA and for one composite, DAPAC5. The cooperative model predictions of the yield stress were in good agreement with the experimental results.

CONCLUSION

This study stressed the very interesting properties of bio-based polyamides from rapeseed oil based-dimer acid (DAPA) reinforced with cellulose fibres.

Tensile tests were carried out for DAPA and DAPA/Cellulose composites. The experimental results showed that the cellulose addition to DAPA polymer matrix led to a high increase Young modulus and yield stress. Good fitting to the experimental data was obtained with the adapted models. In terms of tensile tests results, these biocomposites showed similar behaviour to conventional reinforced thermoplastics, with a high increase of Young modulus and a drastic decrease in the elongation at break. Halpin-Tsai model seems to be an accurate and simple answer to evaluate the elastic modulus evolution. Considering the yield stress, the cooperative model of Richeton showed that the activation energy slightly increased with the cellulose content and the activation volume decreased with increasing cellulose content. As a perspective, the cooperative model used in this work could be implemented in the formulation of large deformation for DAPAC or other cellulose composite materials.

The association of this innovative bio-based polymer with cellulose fillers is a good answer to overcome primary issues with green polymers, i.e. the material cost and the low mechanical properties. Then, this association can be used in different fields such as automobile or coating industries. The key-point is that these materials are in very good agreement with the concept of sustainable development.

REFERENCES

[1] Biermann U, Friedt W, Lang S, Lühs W, Machmüller G, Metzger JO, Rüsch M, Schäfer H Schneider MP. Angew. Chem. - *Int. Ed.* 2000;39:2207-2224.

[2] Guner FS, Yagci Y, *Erciyes AT. Prog. Polym. Sci.* 2006;31:633-670.

[3] Khot SN, Lascala JJ, Can E, Morye SS, Williams GI, Palmese GR, Kusefoglu SH, Wool RP. *J. Appl. Polym. Sci.* 2001;82:703-723.

[4] Meier MAR, Metzger JO, Schubert US. *Chem. Soc. Reviews* 2007;36:1788-1802.

[5] Nayak PL. J. Macromol. S*ci.-Reviews in Macromolecular Chemistry and Physics* 2000;C40:1-21.

[6] Hablot E, Zheng D, Bouquey M, Averous L. Macromol. *Mat. Eng.* 2008;293:922-929.

[7] Cavus S, Gurkaynak MA. *Polym. Adv. Techn.* 2006;17:30-36.

[8] Brandrup J, Immergut EH, Grulke EA, Abe A, Bloch DR. *Polymer Handbook,* 4th Edition Wiley-Interscience. New York: 1999.
[9] Bledzki AK, Gassan J. *Prog. Polym. Sci.* 1999;24:221-274.
[10] Eichhorn SJ, Baillie CA, Zafeiropoulos N, Mwaikambo LY, Ansell MP, Dufresne A, Entwistle KM, Herrera-Franco PJ, Escamilla GC, Groom L, Hughes M, Hill C, Rials TG, Wild PM. *J. Mat. Sci.* 2001;36:2107-2131.
[11] Hablot E, Matadi R, Ahzi S, Avérous L. *Comp. Sci. Tech.* 2010;In press.
[12] Hablot E, Matadi R, Ahzi S, Vaudemond R, Ruch D, Avérous L. *Comp. Sci. Tech.* 2010;In press.
[13] Amash A, Zugenmaier P. *Polymer* 2000;41:1589-1596.
[14] Voigt W. *Lehrbruch des Krystalphysik.* Berlin: 1928.
[15] Reuss A, Angew Z. *Appl. Math. Mechanics* 1929;9:49.
[16] Halpin J, Kardos J. *Polym. Eng. Sci.* 1976;16:344-352.
[17] Gibson R. *Principles of composite material mechanics.* New-York McGraw-Hill, 1994.
[18] Tucker CL, Liang E. *Composites Sci. Tech.* 1999;59:655-671.
[19] Bauwens-Crowet C, Bauwens JC, Homès G. *J. Polym. Sci.: Part A-2* 1969;7:735-742.
[20] Robertson RE. *J. Chem. Phys.* 1966;44:3950-3956.
[21] Roetling JA. *Polymer* 1965;6:311-317.
[22] Buckley CP, Salem DR. *Polymer* 1987;28:69-85.
[23] Llana PG, Boyce MC. *Polymer* 1999;40:6729-6751.
[24] Richeton J, Ahzi S, Daridon L, Remond Y. *Polymer* 2005;46:6035-6043.
[25] Chivrac F, Gueguen O, Pollet E, Ahzi S, Makradi A, Averous L. *Acta Biomaterialia* 2008;4:1707-1714.
[26] Gueguen O, Richeton J, Ahzi S, Makradi A. *Acta Materialia* 2008;56:1650-1655.
[27] Takyanagi M, Uemura S, Minami S. *J. Polym. Sci.: Part C* 1964;5:113-122.
[28] Eyring HJ. *J. Chem. Physics* 1936;4:283.

In: Biodegradable Polymers ...
Editors: A. Jimenez and G. E. Zaikov
ISBN 978-1-61209-520-2
© 2011 Nova Science Publishers, Inc.

Chapter 4

PREPARATION AND PROPERTIES OF THREE LAYER SHEETS BASED ON GELATIN AND POLY(LACTIC ACID)

J. F. Martucci and R. A. Ruseckaite[*]

Research Institute of Material Science and Technology (INTEMA),
J. B. Justo 4302, 7600 - Mar del Plata, Argentina

ABSTRACT

Glycerol plasticized-gelatin (Ge-30Gly) and poly(lactic acid) (PLA) films were prepared by heat-compression molding and then piled together to produce a biodegradable three-layer sheet with PLA as both outer layers and Ge-30Gly as inner layer. Multilayer sheets displayed a compact and uniform microstructure due to the highly compatible individual layers which could interact by strong hydrogen bonding. Lamination reduced total soluble matter compared to the single layers while keeping transparency. Tensile strength of the multilayer sheet (36.2 ±4.3 MPa) increased 16-folds compared with that of Ge-30Gly. Lamination also had beneficial effect on the barrier properties. The Water Vapour Permeation of the multilayer sheet (1.2 ± 0.1 10^{-14} kg·m·Pa^{-1}·s^{-1}·m^{-2}) decreased compared with that of Ge-30Gly while the oxygen permeability (17.1 ± 2.3 cm^3(O2)·mm·m^{-2}·day^{-1}) was reduced with respect

---

[*] E-mail address: ruseckaite@intema.gov.ar

to that of neat PLA and was comparable to the value of Ge-30Gly layer. The individual layers and multilayer sheet were submitted to degradation under indoor soil burial conditions. Biodegradation of multilayer as well as individual components was evaluated by monitoring water absorption and weight loss. During the experiment time, weight loss of multilayer sheet showed two stages. During the first stage, the weight loss of the film increased rapidly due to the loss of Ge-30Gly inner layer. From 25th to 120th day no further changes were observed suggesting that the inner layer was completely biodegraded. Visually, multilayer sheet turned from transparent to opaque after 120 days due to an increment of the crystallinity of the PLA polymer matrix. SEM analysis showed that after 25 days a significant change in the overall morphology of multilayer sheet could be observed. This result suggested that Ge inner layer promote the PLA outer layer degradation, probably because the Ge hydrophilicity allows the water molecules to penetrate more easily within the PLA and to trigger the hydrolytic degradation process faster.

INTRODUCTION

In recent years biopolymers have been visualized as potential environmentally friendly and sustainable alternatives to petroleum based plastics, particularly in those applications where biodegradability and derivatization of natural resources gives added value [1,2]. Gelatin (Ge) is an animal protein, inexpensive and totally biodegradable, with good film-forming ability and excellent oxygen and aroma barrier properties at low relative humidity as well as fairly good mechanical properties appropriate for the production of bio-packaging materials [3]. However, as most biopolymers, the main general resistance to further using gelatin - based films in food packaging is their limited water resistance [4,5,6,7,8,9,10,11,12]. An alternative to overcome this drawback is to combine protein-based films with moisture resistant biodegradable polymer–based films into multilayer films or sheets to exploit the advantages of both polymers, with fewer processing steps [2,4,13]. For this purpose, poly(lactic acid) (PLA) is one of the most promising biopolymers which is obtained from sugar feedstock, corn, etc., renewable resources readily biodegradable [14,15,16]. This thermoplastic aliphatic polyester has high strength, high modulus, and good processability, and it is completely biodegradable. The multilayer film is anticipated to have at least equal performance than that of the commercial laminate films, but with the additional benefit of being biodegradable at the end of film life.

Little information about biodegradable multilayer films or sheets based on proteins and biodegradable polymers is available in the literature. Rhim et al [13,17] reported the preparation of multilayer films based on PLA and plasticized soybean protein isolate (SPI) films. The multilayer resulted in desirable barrier properties and improved mechanical properties compared to those of SPI. However, individual components as well as multilayer films were produced by solvent–casting technique which is time-consuming because some parameters, such as film thickness, are difficult to control [18].

The development of multilayer films by using thermal formation of protein-based materials by techniques common in synthetic thermoplastics processing (extrusion, injection, etc) is the next challenge in an attempt to develop commercial biodegradable packaging films and sheets. Generally, compression molding of sheets is studied as a precursor to extrusion, in order to demonstrate material flowability and fusion and to identify conditions suitable for extrusion [18]. Gelatin can be thermoplastically processed under the plasticization of small molecules, generally polyols [5,19,20]. In a previous work, we developed biodegradable three-layer films based on modified– bovine gelatin produced by compression moulding with improved barrier properties [21]. The outer layers were based on plasticized gelatine cross-linked with dialdehyde starch (DAS) [4] while the inner–gas barrier layer was a bio-nanocomposite film based on plasticized gelatine and 5 wt% of sodium montmorillonite. However, tensile strength of such multilayer films was still insufficient from practical applications. Our current work is focused on developing biodegradable multilayer films and sheets based on gelatin [21] and/or biodegradable polymers. The objective of this work was therefore to produce three-layer sheets based on plasticized gelatin as inner layer and PLA films as outer and moisture resistant layers with potential application to the packaging sector. The performance of the obtained sheets was evaluated in terms of their transparency, moisture resistance, barrier and mechanical properties. The susceptibility to soil microflora was also evaluated through indoor soil burial experiments.

Experimental

Materials

Bovine hide gelatin (Ge) type B was kindly supplied by Rousselot (Argentina), Bloom 150, isoionic point (Ip) 5.3. Poly (d,l-lactic acid) (CML PLA) was purchased from Tate&Lyle (Turkku, Finland) and used as received. Glycerol analytical grade (Gly, 98%), with a molecular weight 92.02 g mol^{-1} was supplied by DEM Chemicals (Mar del Plata, Argentina).

Preparation of Laminated Sheets

Three-layer sheets were prepared through a two-step process. Firstly, individual layers were produced separately by heat-compression molding. PLA pellets were processed into films in a hot press (E.M.S., Buenos Aires, Argentina) at 180 °C. The material was kept between the plates at atmospheric pressure for 5 min and then it was successively pressed under 3 MPa for 1 min, 5 MPa for 1 min and 10 MPa for 3 min [22]. For the preparation of gelatin-plasticized films, gelatin powder was mixed with 30 wt% glycerol (gelatin dry basis) using a kitchen mixer (M.B.Z., San Justo, Buenos Aires, Argentina) at low speed (150 rpm) for 24 h at ambient temperature. The blended Ge-30Gly was transferred into a stainless steel mould (30 x 30 cm^2) and kept between the plates at atmospheric pressure and 120 °C for 5 min. Pressure was risen to 50 kg cm^{-2} for 10 min to obtain homogeneous sheets. Furthermore, samples were kept between plates at atmospheric pressure and then cooled with water up to room temperature [4]. Lamination involved pressing together multilayer films in a press at 100 °C under 50 kg·cm^{-2} for 10 min to obtain PLA/Ge/PLA sheets. The average sheet thickness was 0.45 ± 0.05 mm The obtained specimens were stored under controlled temperature and humidity conditions prior to further tests (65 ± 2% RH; 25 ± 2 °C).

Characterization

Thicknesses of each film and three-layer sheets were measured with a digital micrometer (Vernier, China) with 10-microns resolution. Overall thickness was expressed as the average of five measurements randomly taken around the individual and multilayer sheet testing areas. Opacity was determined by using a UV-Vis spectrophotometer Shimadzu 1601 PC. The absorbance spectrum (400–800 nm) was recorded and the opacity was defined as the area under the recorded curve determined by an integration procedure and it was expressed as Absorbance units (nanometers per thickness unit) (AU·nm·mm^{-1}) [23]. Total soluble matter (TSM) was determined as the percentage of dry matter solubilized after 24 h immersion in distilled water [24,25]. Previously, residual moisture content (RMC) and initial dry matter (%) of the laminated sheet and individual components were gravimetrically determined (± 0.0001 g) by drying samples at 105 °C in an oven with forced air circulation for 24 h. TSM results were the mean values of five independent measurements. Water Vapour Permeability (WVP, kg.m m^{-2}.Pa^{-1}.s^{-1}) was gravimetrically determined by following ASTM E96-95 standard [26]. Values were the average of five replicates. Oxygen Permeation (OP) was measured with an Oxygen Permeation Analyzer from Systech Instruments, model 850. The gas volumetric flow rate per unit area of the membrane, OTR (cm^3 (O$_2$)·day^{-1}·m^{-2}) was continuously monitored until a steady state was reached (OTR∞). For a constant partial pressure gradient across the polymer film throughout all experiments, OP is proportional to OTR∞·e (cm^3(O$_2$)·mm·day^{-1}·m^{-2}) where e is the specimen thickness (mm). Tensile properties were evaluated according to ASTM D638-94b standard [27] by using an Instron Universal Testing Machine equipped with a 0.5 KN cell and at a crosshead speed of 3 mm min^{-1}. Seven specimens of each film were analyzed and data for each test were statistically treated.

Indoor soil degradation experiments were carried out in a series of plastic boxes, containing characterized soil (Pinocha type = typical Argentinean Argiudol + pine litter) and the natural microflora present was used as degradation medium. This soil showed pH 6.1 (H$_2$O), organic matter content (OMC), 7.2%; total nitrogen, 0.19%; NO$_3^-$, 18ppm; P, 10.5 ppm; Ca, 12.5 meq/100g; Mg, 2.1 meq/100g ; Na, 0.3 meq/100g; K, 1.5 meq/100g [28]. The relative humidity was kept around 40% by adding water periodically and the temperature was set at 20 ± 2°C. After a predetermined degradation time, polymer specimens were carefully removed from the soil to avoid damage and thoroughly cleaned with distilled water to determine water absorption (WS). Samples were removed from the soil at specific intervals (t), carefully cleansed with distilled water, superficially dried with a tissue paper and further weighed (m_h). Water uptake (%WS) was quantified by using the following equation:

$$\%WS = \frac{m_h - m_t}{m_0} * 100 \quad (1)$$

where m_0 and m_t are the initial and the residual mass at time = t, respectively, and m_h is the humid mass of the specimens after wipping with a tissue paper. The values reported are the average of three measurements.

After water sorption determination, samples were dried under vacuum at room temperature to constant weight. The specimens were weighed on an analytical balance in order to determine the average weight loss (%WL):

$$\%WL = \frac{m_t - m_0}{m_0} * 100 \quad (2)$$

where m_0 is the initial mass, m_t is the remaining mass (after drying) after a time = t of incubation. All results are the average of three replicates.

All experimental data were analyzed and compared by using the analysis of variance (ANOVA) ($\alpha = 0.05$). Comparison of means was carried out by using a Tukey test to identify which groups were significantly different from other groups (P<0.05).

RESULTS AND DISCUSSION

Three-layer sheets were found to be homogeneous, without holes and cracks, with good visual aspect and transparency (Figure 1). Optical properties of Ge-30Gly, PLA and multilayer sheets are presented in Table 1. The obtained results indicated that Ge-30Gly had the highest opacity value (490 ± 54 UA·nm·mm^{-1}) which was in accordance to values previously reported for plasticized gelatin films [4,29]. Interestingly, the opacity value of the multilayer sheet did not differ significantly (P>0.05) from the value measured for neat PLA film. Since thickness values of the individual components did not differ significantly (P>0.05), the reduction in opacity of Ge-30Gly films when laminating with PLA could reflect a good compatibility between layers [13,30]. If interfaces between Ge-30Gly and PLA exist, when light reaches these interfaces, light scattering should occur and transmission should be reduced, increasing consequently the sample opacity. In addition, manual peeling of PLA layers from the inner Ge-30Gly was not possible. It could be concluded that thermal compression molding favored the bond strength between PLA and gelatin layers by hydrogen interactions through the carbonyl groups from PLA and hydrogen from peptide bonds in gelatin or hydroxyl groups in glycerol. Rhim et al. (2006) [13] proposed the presence of such kind of interactions in multilayer films based on PLA and SPI produced by casting.

High water resistance of a film is one of the most important properties for food packaging, especially for high water activity foods or foods coming into contact with high-humidity environments during transportation and storage. Total soluble matter (TSM) of individual and multilayer sheets was determined to evaluate the integrity of such materials in aqueous environment (Table 1). While Ge-30Gly film was completely soluble in water [4], PLA film was insoluble at least during the time of the study (24 h). The low TSM value of the multilayer was attributed to the solubilization of the Ge-30Gly inner layer indicating that hydrophobic PLA outer layers are the main factor responsible of the integrity of the multilayer during soaking [13].

Water vapor permeability (WVP) and oxygen transmission rate (OTR) are two of the main barrier properties studied in packaging applications [31]. Multilayer sheet exhibited the lowest WVP value (Table 1). WVP of the multilayer improved 10-times compared to Ge-30Gly film and was similar to that of PLA. This was associated to the hydrophobic character of PLA [17]. Lamination reduced significantly the OTR∞·e value of PLA. It is well known that proteins are good gas barriers at low and medium relative humidity [4,5]. Therefore, the improvement in OTR could be caused by the oxygen barrier role of the inner Ge-30Gly film [4,32]. OTR∞·e results of the three-layer sheet were still lower than those obtained for LDPE in similar conditions (OTR∞·e$_{LDPE}$ = 160cm^3·mm·m^{-2}·day^{-1}). As this material is currently used in films manufacturing, the use of multilayer sheet could be acceptable for food packaging with reduced oxygen permeation.

Figure 1. Visual appearance of Ge-30Gly, PLA and multilayer sheet.

Table 1. Opacity, Total Soluble Matter (TSM), Water Vapor Permeability (WVP), Oxygen Transmission Rate per thickness (OTR∞·e) of individual films and multilayer sheet

Sample name	Opacity (UA·nm·m^{-1})	Thickness (mm)	TSM 24hs. (%)	WVP*10^{14} (Kg·m·Pa^{-1}·s·m^{-2})	OTR∞·e (cm^3·mm· m^{-2}·day^{-1})
Ge-30Gly	490 ± 54^a	0.43 ± 0.02^a	100.0 ± 0.0^a	13.6 ± 3.9^a	13.3 ± 3.0^a
PLA	258 ± 13^b	0.42 ± 0.03^a	0.02 ± 0.04^b	1.3 ± 0.1^b	29.5 ± 6.4^b
PLA/Ge/PLA	344 ± 57^b	0.46 ± 0.05^a	8.99 ± 1.53^c	1.2 ± 0.1^b	17.1 ± 2.3^a

Note: Any two means in the same column followed by the same letter are not significantly (P>0.05) different according to Turkey test.

Mechanical properties of multilayer and control films are summarized in Table 2. Tensile strength (TS) and elongation at break (ε_B) of Ge-30Gly film were 2.31 ± 0.59 MPa and 124.0±26.8%, respectively. The low TS value of Ge-based films restricts their use in food packaging applications and consequently this property should be improved i.e. by laminating with high TS biopolymers, such as PLA, with high amorphous degree and behaving as a brittle material with high modulus (around 2,0 GPa), tensile strength (45 MPa) and low deformation at break (about 9%) [22]. On the other hand, TS and E values of the multilayer sheet were significantly ($P < 0.05$) increased compared to those of Ge-30Gly films. Tensile strength value of multilayer sheet (36.2 ± 4.3 MPa) is similar to HDPE (26 MPa) [33]. The enhancement in E and TS (7.76 ± 0.75 % respectively) are an indirect evidence for the compatibility of Ge and PLA in multilayer sheet [13].

Table 2. Young Modulus (E), Tensile strength (TS) and Elongation at break (EB) of films

Sample name	Mechanical Properties		
	E(MPa)	EB (%)	TS (MPa)
Ge-30Gly	7.4 ± 1.8^a	124 ± 27^a	2.3 ± 0.6^a
PLA	2073 ± 174^b	8.6 ± 0.9^b	45.0 ± 1.4^b
PLA/Ge/PLA	1583 ± 189^c	7.8 ± 0.7^b	36.2 ± 4.3^c

Note: Any two means in the same column followed by the same letter are not significantly (P>0.05) different according to Turkey test.

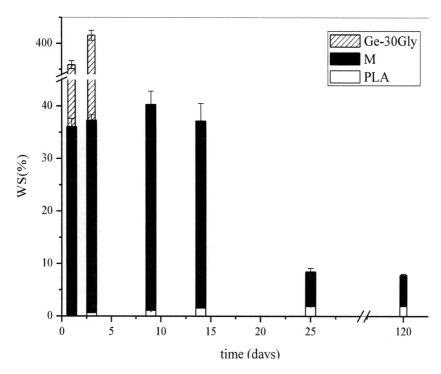

Figure 2. Water absorption (WS) of Ge, PLA and multilayer sheets after different exposition times to degrading medium.

In order to evaluate the susceptibility of three-layer sheets to aerobic degradation, control films and multilayer sheet were exposed to natural microbial consortia as degrading environment. Gelatin is reported to be rapidly degraded when the contamination and the environmental conditions are adequate. Bacteria from the genera *Bacillus* and *Pseudomonas,* as well as the yeast-like *Aureobasidium sp.* have been reported as effective in degrading gelatin. Filamentous fungi (i.e. *Aspergillus*) has also been found to be active even at low temperatures such as 4 °C [34,35]. On the other hand, the degradation of PLA and oligomers has been investigated in vitro and outdoor conditions [36,37]. PLA has been classified as a bio-assimilable polymer due to the degradation mechanism; that is, at the beginning of the degradation process, the hydrolytic degradation leads to oligomeric lactic acid products, which can be further bio-assimilated or mineralized by microorganisms such as fungi or bacteria.

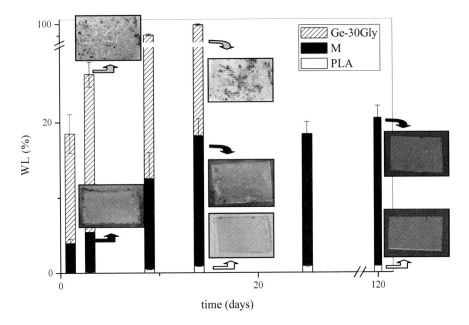

Figure 3. Weight loss (WL) and optical photographs of individual Ge-30Gly and PLA films and multilayer sheet after different exposition time to soil burial.

The progressive percentage of weight gain of these materials due to water absorption during soil burial is shown in Figure 4. WS of control Ge-30Gly was very fast attaining 400% within the first 2 days due to gelatin hydrophilic character and the hydrophilic nature of glycerol. Beyond the 2nd day of soil burial, water absorption dropped slightly as immersion time was increased, owing to the fact that some low-molecular chain peptides from gelatin were leached away from the specimen. Therefore, estimation of water absorption for Ge-30Gly is difficult and could lead to wrong results because gravimetrical measurement are prone to errors due to weight loss induced by degradation processes [2].

In contrast, PLA appeared more hydrophobic; WS slowly increased to reach only 4% after 120 days of exposure. The low amount of moisture absorbed by neat PLA agreed well with its hydrophobic nature. Similar results were previously reported for PLA50 and PLA100 in soil [36] and PLA/starch bars buried in soil [38]. The WS of the multilayer was initially faster than the result for neat PLA, reaching a maximum of 42% after 10 days. Beyond the 14th day, WS declined. This may be attributed to the penetration of water molecules into the interfacial zone i.e. between PLA/Ge-30Gly/PLA. The inner Ge-30Gly layer is a hydrophilic material that could enhance H-bonds formation with water which penetrates through the un-protected edges of sample. Thus, the presence of moisture at the PLA/Ge-30Gly interface further weakened the adhesion of the individual layers and delamination eventually occurred. Thereafter, water absorption leveled off to reach 8% after 120 days, similar to values observed for neat PLA. From these results, it appeared that hydrophilic Ge-30Gly inner layer and interfacial interactions between layers govern the moisture uptake of the multilayer during al least the first stages of the process.

Weight loss pattern of multilayer sheet and individual films is shown in Figure 3 where their mass, normalized with respect to its initial mass, is plotted versus exposition time. The main difficulties with this method are either loss of sample fragments as a result of the thorough cleaning process or an increase in sample weight due to leftover soil debris. Therefore values shown were the average of three replicates. The maximum WL achieved showed the following tendency PLA < multilayer < Ge-30Gly. These results are more likely related to the water absorption capacity of the different films. Ge-30Gly control film exhibited a sustained WL until reaching 100% the 14th day. Gelatin may suffer abiotic and enzymatic hydrolysis due to the action of enzymes (proteases) which are present in a variety of microorganisms [2,39]. Gelatin is also sensitive to the attack of filamentous fungi under humid conditions [2,39,40]. On the other hand, neat PLA showed slight weight change over the 120 days of the study. Such behavior is not surprising since PLA biodegrades slowly in soils at ambient temperature because of the slow rate of hydrolysis at temperatures below the glass transition temperature ($T_{g\ PLA}$ = 66 °C [22]) and under reduced humidity conditions [41,42]. Indeed, moisture absorption during degradation plays a very important role since according to Torres et al [36], PLA plates must be initially degraded by chemical hydrolysis until the formation of by-products (short chain PLA, cyclic oligomers) which are able to be processed via biochemical routes by microorganisms and indirectly by animals of the food chain like earthworms [43]. The multilayer showed a rapid WL during the first 14 days with a maximum about 18%. Thereafter the WL value leveled off reaching a final value of 20% after 120 days. The higher WS of the multilayer when compared to neat PLA could facilitate the abiotic hydrolysis and subsequent bio-assimilation of PLA, contributing to the slight increment in WL in the later stages of the study.

Changes in the general aspect of samples upon soil burial could be also visually examined. Multilayer sheet increased its relative opacity with incubation time (Fig. 3). Despite the inherent amorphous character of the PLA used in the present study this increment in opacity (or whitening) could be ascribed to the so-called "degradation-induced crystallization" previously reported [44,45]. The appearance of crystallinity was quantified by XRD analysis performed on multilayer degraded samples (Table 3). Before biodegradation, PLA in the multilayer was essentially amorphous, but crystallinity increased throughout the experiment. This behavior could be reasonably attributable to the higher water uptake of the multilayer sheet induced by the Ge-30Gly inner layer. This would promote the chain break into shorter dimensions and would give mobility to polymer chains and fragments resulting in a re-arrangement into a more organized structure. In addition, water intake promotes the chain break. In contrast, for neat PLA film, crystallization was only observed at later stages of degradation in agreement with its low moisture uptake measured during incubation.

Table 3. Crystallinity of PLA and multilayer sheets at different exposition time to soil burial

Incubation time (days)	Cristallinity (%)	
	PLA	Multilayer
0	0	9.3
7	0	26.4
14	0	27.4
40	0	33.6
120	23.0	52.8

CONCLUSIONS

Three-layer sheets with PLA as outer layers and glycerol-plasticized gelatin as inner layer were produced by thermal compression molding without compatibilizing agents, adhesives or chemical modification of film surfaces. The synergetic effect of lamination was evidenced by the increased tensile strength (16-folds), Young modulus (214-folds), WVP (92%) and oxygen barrier (40%) as well as impact properties of neat PLA. The combination of the mechanical strength and hydrophobicity of PLA with the gas barrier properties of plasticized gelatin film might lead to a multilayer sheet with appropriate properties for food packaging. Since both materials are environmentally sound, the three-layer sheet is anticipated to be biodegradable. Indoor soil burial test has demonstrated that the multilayer film is less resistant to the attack by microorganisms than PLA control film probably because the Ge inner layer allows the water molecules to penetrate more easily within the PLA and to prompt the hydrolytic degradation process faster.

AKCNOWLEDGMENTS

Authors want to express their gratitude to the National Research Council (CONICET, Argentina), contract grant number: PIP 112-200801-01837 and to SECyT (APCyT, Argentina); contract grant number: PICT06-1560, for their financial support.

REFERENCES

[1] Tromp OS. *Sustainable use of renewable resources for material purposes,* 1995. UNEP-WG-SPD, Amsterdam.
[2] Martucci JF, Ruseckaite RA. *Polym. Degrad. Stabil.* 2009;94:1307-1313.
[3] Jongjareonrak A, Benjakul S, Visessanguan W, Prodpran T, Tanaka M. *Food Hydrocolloid.* 2006;20:492-501.
[4] Martucci JF, Ruseckaite RA. *J. Appl. Polym. Sci.* 2009;112:2166-2178.
[5] Gennadios A. *Protein-based films and coatings.* 1st. Ed., CRC Press, pp 1-41, (2002).
[6] Arvanitoyannis I, Nakayama A. Aiba S.I. *Carbohyd. Polym.* 1998;36:105-119.
[7] Apostolov AA, Fakirov S, Evstatiev M, Hoffman J, Friedrich K. *Macromol. Mater. Eng.* 2002;287:693-697.
[8] de Carvalho RA, Grosso CRF. *Food Hidrocolloid.* 2004;18:717-726.
[9] Bigi A, Cojazzi G, Panzavolta S, Roveri N, Rubini K. *Biomaterials* 2002;23:4827-4832.
[10] Bertan LC, Tanada-Palmu PS, Siani AC, Grosso CRF. *Food Hydrocolloid.* 2005;19:73-82.
[11] Avena-Bustillos RJ, Olsen CW, Olson DA, Chiou B, Yee E, Bechtel PJ, Mchugh TH. *J. Food Sci.* 2006;71:E202-E207.
[12] Arnesen JA, Gildberg A. *Bioresource Technol.* 2007;98:53-57.

[13] Rhim JW, Mohanty KA, Singh SP, Ng PKW. *Ind. Eng. Chem. Res.* 2006;45:3059-3066.
[14] Weber CJ, Haugaard V, Festersen R, Bertelsen G. *Food Addit. Contam.* 2002;19:172-177.
[15] Roper H. Starch/Starke 2002;54:89-95.
[16] Cabedo L, Feijoo JL, Villanueva MP, Lagarón JM, Giménez E. *Macromolecular Symposia* 2006;233:191-197.
[17] Rhim JW, Lee JH, Ng PKW. LWT – *Food. Sci. Technol.* 2007;40:232-238.
[18] Sothornvit R, Olsen CW, McHugh TH, Krochta JM. *J. Food Eng.* 2007;78:855-860.
[19] Kumar R, Choudhary V, Mishra S, Varma IK, Mattiason B. *Ind. Crop. Prod.* 2002;16:155-172.
[20] Park JW, Whiteside WS, Cho SY LWT - *Food Sci. Tech.* 2008;41:692-700.
[21] Martucci JF, Ruseckaite RA. *J. Food Eng.* In press, (2010).
[22] Martino VP, Jiménez A, Ruseckaite RA. *J. Appl. Polym. Sci.* 2009;112:2010-2018.
[23] Irissin-Mangata J, Bauduin G, Boutevin B, Gontard N. Eur. *Polym. J.* 2001;37:1533-1541.
[24] Guilbert S, Gontard N, Cuq B. Packag. *Technol. Sci.* 1995;8:339-346.
[25] Rhim JW, Gennadios A, Weller CL, Cezeirat C, Hanna MA. *Ind. Crop Prod.* 1998;8:195-203.
[26] American Society for testing and Materials. Standard Test Methods for Water Vapor Transmission of Materials, *Desiccant Method,* E96-95 Philadelphia, Pa: ASTM, 1995.
[27] American Society for Testing and Materials. *Standard test method for tensile properties of plastics.* D638-94b. Philadelphia, Pa: ASTM, 1995.
[28] Martucci JF. Structure-properties relationship in materials based on gelatin. *Doctoral Thesis*, Mar del Plata University, Argentina, March 2008.
[29] Rivero S, García MA, Pinotti A. *J. Food Eng.* **2009;90:**5311-5319.
[30] Martucci JF, Ruseckaite RA, Vázquez A. *Mater. Sci. Eng.* A 2006;435-436:681-686.
[31] Siracusa V, Rocculi P, Romani S, Rosa MD *Trends Food Sci. Technol.* 2008;19:634-643.
[32] Fang JM, Fowler PA, Escrig C, González R, Costa JA, Chamudis L.*Carbohyd. Polym.* 2005;60:39-42.

[33] Salamie M. Polyethylene, low density. In M. Bakker (Ed.), *The Wiley encyclopedia of packaging technology,* 514–523. New York:Wiley. (1986).
[34] Abrusci C, Martín-González A, Del Amo A, Corrales T, Catalina F. *Polym. Degrad. Stabil.* 2004;86:283–291.
[35] Abrusci C, Martín-González A, Del Amo A, Catalina F, Collado J, Platas G. *Int. Biodeterior. Biodegrad.* 2005;56:58–68.
[36] Torres A, Li SM, Roussos S, Vert M. *Appl. Environ. Microb.* 1996;62:2393–2397.
[37] Torres A, Li SM, Roussos S, Vert M. Microbial degradation of poly(lactic acid)as a model of synthetic polymer degradation mechanisms in outdoor conditions, *Biopolymer chapter 14,* 218-226. ACS Symposium Series vol. 723, (1999).
[38] Gattin R, Copinet A, Bertrand C, Couturier Y. *Int. Biodeterior. Biodegrad.* 2002;50:25-31.
[39] Dalev PG, Patil RD, Mark JE, Vassileva E, Fakirov S. *J. Appl. Polym. Sci.* 2000;78:1341-1347.
[40] Abrusci C, Marquina D, Santos A, Del Amo A, Corrales T, Catalina FA. *J. Photochem. Photobiol. A.* 2007;185:188-197.
[41] Pranamuda H, Tsuchii A, Tokiwa Y. *Macromol. Biosci.* 2001;1:25-29.
[42] Shogren RL, Doane WM, Garlotta D, Lawton JW, Willett JL. *Polym. Degrad. Stabil.* 2003;79:405-411.
[43] Alauzet N, Garreau H, Bouche M, Vert, M. *J. Polym. Environ.* 2002;10:53-58.
[44] González MF, Ruseckaite RA, Cuadrado TR. *J. Appl. Polym. Sci.* 1999;71:1223-1230.
[45] Migliaresi C, Fambri L, Cohn D. *J. Biomater. Sci. Ed.* 1994;5:591-606.

In: Biodegradable Polymers ...
Editors: A. Jimenez and G. E. Zaikov

ISBN 978-1-61209-520-2
© 2011 Nova Science Publishers, Inc.

Chapter 5

EVALUATION OF THE USE OF NATURAL PLASTICIZERS IN COMMERCIAL LIDS FOR FOOD PACKAGING. CHARACTERIZATION AND MIGRATION IN FOOD SIMULANTS

C. Bueno-Ferrer[*], M. C. Garrigós and A. Jiménez

University of Alicante. Analytical Chemistry, Nutrition and Food Sciences Department. PO Box 99, 03080, Alicante, Spain

ABSTRACT

Epoxidized soybean oil (ESBO) is a vegetable oil widely used as plasticizer and/or stabilizer for poly (vinyl chloride) (PVC) formulations in food contact materials, in particular in gaskets of lids for glass jars. ESBO shows some advantages, such as low toxicity to humans and biodegradable nature, compared to common additives for PVC, making very attractive its use in food packaging. However, its migration to foodstuff is a crucial issue for this application. The current specific migration limits (SME) established for ESBO in food contact materials make necessary the study of its presence and migration to food simulants, since the tolerable daily intake (TDI) of 1 mg kg^{-1} body weight is often exceeded.. In this work, a wide screening of the use of ESBO in

---

[*] E-mail: carmen.bueno@ua.es; mc.garrigos@ua.es; alfjimenez@ua.es

commercial lids for glass jars was carried out and this additive was identified and determined in most of the commercial samples tested. ESBO migration to food simulants was also determined for controlling compliance of the current legal migration limits. Thermal characterization of commercial lids was also carried out in order to explain differences in composition.

INTRODUCTION

The use of natural products, in particular those obtained from vegetal oils and fats, can be considered an adequate and environmentally-friendly alternative to phthalates in plasticized poly(vinyl chloride) (PVC) [1]. Epoxidized soybean oil (ESBO) is a vegetable oil widely used as plasticizer and/or stabilizer for PVC formulations used in food contact materials [2-6]. It was reported that the ESBO stabilization mechanism proceeds through a reaction between the epoxide ring from ESBO and hydrogen chloride generated during PVC degradation, restoring the labile chlorine atoms back into the polymer chains [7]. This reaction prevents PVC from further dehydrochlorination, preserving its colour and limiting loss in properties at high temperatures.

This material has shown some applications in food packaging, such as gaskets of lids for glass jars or in domestic films where PVC usually contains 35-45 wt% of plasticizer [5,6,8,9]. ESBO, as well as other lower molar mass plasticizers such as acetyl tributyl citrate (ATBC), dioctyl adipate (DOA) or dibutyl sebacate (DBS), has grown during the last years as a valid alternative to phthalates, under discussion because of their potential toxicity to humans [10].

ESBO shows relatively good compatibility with PVC, low toxicity to humans [11-13] and biodegradable nature [14-18]. However, migration to foodstuff is still under study, in particular in high-fat foods [3,5,8,19]. The specific migration limit (SML) established for ESBO in food contact materials is 60 mg kg^{-1} food simulant [20]. However, specific regulation (2005/79/EC) [21] lowered the ESBO SML to 30 mg kg^{-1} for materials in contact with infant food. Then, a transition period was agreed in order to give industry more time to find alternatives. Therefore, ESBO migration limit was increased to 300 mg kg^{-1} food simulant (except for infant food) (EU Regulations 372/2007 and 597/2008) [22,23] until 30 April 2009 but this limit is still in force since it has not been updated.

The aim of this work was to screen the presence and quantity of ESBO in commercial lids of glass jars and its migration to food simulants for controlling compliance of the current legal migration limits, as well as a thermal characterization of these commercial lids.

EXPERIMENTAL

Method for Standard Preparation and Determination of Gasket Composition

ESBO standards were obtained by transmethylation and synthesis of 1,3-dioxolane derivatives as reported elsewhere [24]. The obtained stock solution (65000 mg kg^{-1}) was stored at -18 °C. ESBO calibration standards for GC/MS analysis were prepared by appropriately dissolving an aliquote of this solution in n-hexane. Composition of each gasket was studied by performing Soxhlet extraction with n-hexane for 5 hours (in triplicate) with samples weighing approximately 1g and cut in small pieces. Then, solvent was evaporated from Soxhlet extracts and derivatized as standards.

Unlike ESBO, plasticizers such ATBC, DOA and DBS do not need previous derivatization for GC/MS analysis, due to their lower molar mass. A multi-standard plasticizer solution (1000mg kg^{-1}) in n-hexane was prepared for the migration study in commercial lids. Composition of each gasket was studied by Soxhlet extraction in n-hexane for 5 hours (in triplicate) and further analysis by GC/MS.

Method for Specific Migration Tests from Commercial Lids

Specific migration tests from commercial lids were performed in triplicate by following EN 13130-1 Standard [25]. Glass jars were filled with the appropriate food simulant according to their pH value. Distilled water was selected for asparagus and an aqueous solution (3%w/v) of acetic acid for pickled gherkins, jam and three different fruit-based baby foods. For jars with fat foods (pate, pesto sauce and mayonnaise), ethanol (95%v/v) and iso-octane were selected as alternative fat simulants. A blank test for each simulant was also performed. Glass jars were closed with their lid and turned on its head during 10 days at 40 °C for contact with the respective simulants, except for those samples with iso-octane which were exposed to 2 days at 20 °C, as it is specified in the mentioned Standard [25]. All simulants were subsequently evaporated to dryness in a rotary evaporator.

In order to quantify ESBO migration from gaskets, the derivatization process described in [24] was applied to dried samples before GC/MS analysis. For determination of ATBC, DOA and DBS, dried samples were simply diluted in n-hexane prior to their injection in the gas chromatograph.

GC/MS Analysis

GC/MS analysis was carried out by using an HP/Agilent Technologies 6890N (Palo Alto, CA, USA) gas chromatograph coupled to an Agilent Technologies 5973N mass spectrometer operating in electronic impact (EI) ionization mode (70 eV). A DB-5MS capillary column (30 m x 0.25 mm I.D. x 0.25 µm film thickness) (Teknokroma, Barcelona, Spain) and a split-splitless injector were used. Helium was used as the carrier gas with 1 mL min^{-1} flow rate. Temperatures for injector and detector were 290 °C and 300 °C, respectively. For ESBO determinations, the column temperature was programmed from 140 °C (hold 2 min) to 300 °C at 20 °C min^{-1} heating rate (hold 10 min). Samples (1 µL) were injected in the split mode (split ratio 1:40). Determination of derivatized epoxy fatty acids was performed by using selected ion monitoring (SIM) mode focused on m/z 309, by comparing chromatographic peak areas for migration extracts with those of standards in the same concentration range. Other ions used for identification were those with m/z 277, 367, 396, 465 and 494.

For ATBC, DOA and DBS determination, the column temperature was programmed from 80 °C to 160 °C at 30 °C min^{-1} heating rate and from 160 °C to 320 °C at 15 °C min^{-1} heating rate (hold 7 min). Samples (1 µL) were injected in the split mode (split ratio 1:25) and quantification of plasticizers was performed by using scan mode (50-550 m/z).

Thermogravimetric Analysis (TGA)

Dynamic TGA tests were carried out by using TGA/SDTA 851e Mettler Toledo (Schwarzenbach, Switzerland) equipment. Gaskets from commercial lids were analyzed by weighing approximately 7mg of each one in alumina pans (70µL) and they were heated from 30 °C to 700 °C at 10 °C min^{-1} under nitrogen atmosphere (flow rate 30 mL min^{-1}).

Differential Scanning Calorimetry (DSC)

Tests were performed in TA Instruments Q2000 (New Castle, DE, USA) equipment. Gaskets from commercial lids were analyzed by weighing approximately 3mg of each one in aluminium pans (40µL) and they were subjected to two heating-cooling cycles from -90 °C to 150 °C at 10 °C min^{-1} under nitrogen (flow rate 50 mL min^{-1}).

RESULTS AND DISCUSSION

Gasket Composition

Gaskets composition was determined by GC/MS analysis with previous Soxhlet extraction. The already reported increased use of ESBO as alternative to phthalates in food packaging [2-9] was confirmed in this study since ESBO was found in all gaskets in different amounts ranging from 0.2 to 14.1 wt% (Figure 1). Furthermore, some other plasticizers such as ATBC, DOA and DBS also appeared in some cases. On the other hand, it could be noticed that all plasticizers were detected at lower levels than those expected by considering the normal uses in PVC industry. In this sense, it could be supposed that other compounds, in particular polymeric plasticizers not directly detectable by GC/MS, could be used in these gaskets as co-plasticizers [20,25] in order to complete the plasticization range of gaskets and also to fulfil specific legal migration limits.

Specific Migration Study in Commercial Gaskets

Nine different commercial lids were selected for the specific migration analysis and a screening study of other plasticizers commonly used in gaskets [9] was carried out in order to evaluate the current compliance level of migration in food contact materials in the Spanish market.

The obtained results (Table 1) revealed that the highest amounts of migrated plasticizers were found in alternative fat simulants, particularly in ethanol 95% (v/v). Some plasticizers were found in the three analyzed lids from fat food, all of them above the current legal migration limits. Two of them showed high levels of ESBO and, in the case of pate lid, even above the current transitional limit of 300 mg kg^{-1}. The third one (pesto sauce) showed high levels of ATBC. In the case of aqueous simulants, migration from lids was barely detected and no detectable values were obtained for baby fruit lids.

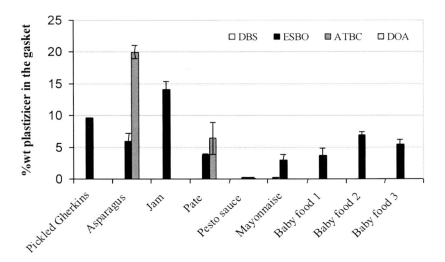

Figure 1. Plasticizers found by GC/MS in gaskets of commercial lids.

Thermogravimetric Analysis (TGA)

Thermal stability of commercial lids was studied by TGA. Table 2 summarizes the results for the initial degradation temperatures (T$_{95\%}$) obtained for all materials. Values of T$_{95\%}$ did not show large differences between most of the gaskets, except in the case of both asparagus and pesto sauce gaskets, where T$_{95\%}$ values were clearly lower to all other materials. The explanation of this behaviour could be obtained when observing GC/MS results for plasticizers composition (Figure 1).

Table 1. Migration of plasticizers from commercial lids to food simulants

Plasticizer	Mean value (mg Kg⁻¹ food simulant)[a]							
	Pick-led Gherkins	Aspa-ragus	Jam	Pate		Pesto sauce		Mayonnaise
ESBO LOQ: 15.9 (mg/kg)	<LOQ	ND	<LOQ	676±282	94±49	18±18	<LOQ	71±24	<LOQ
DBS LOQ: 0.006 (mg/kg)	ND	ND	ND	ND	ND	ND	ND	2±2	0.4±0.2
ATBC LOQ: 0.01(mg/kg)	ND	1.2±0.1	ND	22±19	8±2	285±12	19±3	ND	ND
DOA LOQ: 0.01 (mg/kg)	ND	ND	ND	ND	0.23±0.05	ND	0.23±0.01	0.23±0.01	0.53±0.04
ND: non detectable	Alternative fat simulant: ethanol 95% v/v				Alternative fat simulant: iso-octane			

Note: [a] three replicates.

Table 2. Initial degradation temperatures ($T_{95\%}$, °C) of gaskets from commercial lids determined by TGA analysis at α = 5%

Real samples (Gaskets)	$T_{95\%}$ (°C)
Pickled Gherkins	268
Asparagus	249
Jam	267
Pate	261
Pesto sauce	242
Mayonnaise	267
Baby food 1	258
Baby food 2	258
Baby food 3	260

Pesto sauce gaskets showed no ESBO at all in their composition, and consequently no stabilization from this compound should be expected. The case of asparagus gaskets is a bit more complicated, since some ESBO (around 6 ± 1 wt%) was determined in their composition. However, the main difference in composition between asparagus and the other gaskets was the presence of an important amount of ATBC (around 20 wt%) in their composition. This plasticizer has shown far lower values for $T_{95\%}$ (around 145 °C) than any other plasticizer detected in gaskets. Therefore, a double effect could be expected. On one side the stabilizing effect by ESBO and on the other the reduction in $T_{95\%}$ by the major amount of ATBC in these compositions.

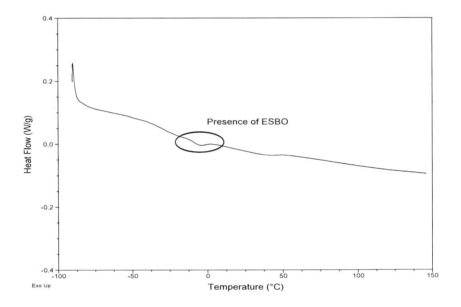

Figure 2. DSC curve of the gasket from jam jar.

Differential Scanning Calorimetry (DSC)

DSC analysis of gaskets showed clearly the presence of ESBO only in the gasket from the jam jar (Figure 2), which had the highest level of ESBO (about 14 wt%). The characteristic crystalline polymorphism of ESBO could be noted in the peak at -10.3°C.

CONCLUSIONS

It was observed that ESBO is used as a common plasticizer in gaskets of lids of glass jars in food packaging systems currently commercialized in Spain, although all the analyzed gaskets contained also another plasticizer meaning that ESBO is mainly used as co-plasticizer in flexible PVC formulations, which could be a reasonable practice used by industry in order to fulfil specific migration limits for each plasticizer. Migration studies from commercial lids confirmed that the amounts obtained for ESBO exceeded legal migration limits in two of the three analyzed lids from fat foods, with the third one exceeding the ATBC legal migration limit.The use of ESBO as additive in flexible PVC formulations should be carefully revised in order to avoid any problem caused by migration in lids from glass jars.

REFERENCES

[1] Demertzis PG, Riganakos KA, Akrida-Demertzi K. *Eur Polym J* 1991;27:231-235.
[2] Lau O, Wong WS. *J Chromatogr A*. 2000;882:255-270.
[3] Fankhauser-Noti A, Fiselier K, Biedermann S, Grob K. *J Food Chem. Toxicol*. 2006;44:1279-1286.
[4] Duffy E, Gibney M. *J Food Addit Contam*. 2007;24(2):216-225.
[5] Biedermann M, Fiselier K, Grob K. *Trends Food Sci Technol*. 2008;19:145-155.
[6] Pedersen GA, Jensen LK, Fankhauser A, Biedermann S, Petersen JH, Fabech B. *Food Addit Contam*. 2008;25(4):503-510.
[7] Parreira TF, Ferreira MMC, Sales HJS., de Almeida WB. *Soc Appl Spectroscopy* 2002;56:1607-1614.
[8] Fankhauser-Noti A, Fiselier K, Biedermann S, Biedermann M, Grob K, Armellini F, Rieger K., Skjevrak I. *Eur Food Res Technol*. 2005;221:416-422.
[9] Fankhauser-Noti A, Grob K. *Trends Food Sci Technol*. 2006;17:105-112.
[10] Rhee GS, Kim SH, Kim SS, Sohn, KH, Kwack SJ, Kim BH, Park KL. *J Toxicol In Vitro* 2002;16:443-448.
[11] BIBRA. British Industrial Biological Research Association Industrial (BIBRA): Toxicity profile: epoxidised soybean oil. 1988, Carlshalton, Surrey, UK.
[12] European Chemicals Bureau (ECB). IUCLID (International Uniform Chemical Information Database) for epoxidized soybean oil. 2000. Available from: http://ecb.jrc.ec.europa.eu/IUCLID-DataSheets/8013078.pdf

[13] European Food Safety Authority (EFSA). Opinion of the Scientific Panel on Food Additives, Flavourings, Processing Aid and Materials in Contact with Food (AFC) on a request from the Commission related to the use of Epoxidized soybean oil in food contact materials adopted on 26 May 2004. EFSA J. 2004;64:1. Available from: http://www.efsa.europa.eu/cs/BlobServer/Scientific_Opinion/opinion_afc10_ej64_epox_soyoil_en1.pdf?ssbinary=true
[14] Choi JS, Park WH. *Polym Testing.* 2004;23(4):455-460.
[15] Liu Z, Erhan SZ, Akin DE, Barton FE. *J Agric Food Chem.* 2006;54:2134-2137.
[16] Behera D, Banthia AK. *J Appl Polym Sci.* 2008;109:2583-2590.
[17] Bouchareb B, Benaniba MT. *J Appl Polym Sci.* 2008;107:3442-3450.
[18] Liu Z, Erhan SZ. *J Mater Sci Eng.* 2008;483-484:708-711.
[19] Grob K, Marmiroli G. *Food Control.* 2009;20(5):491-500.
[20] European Commission (EC). 2002. Commission Directive 2002/72/EC of 6 August 2002 relating to plastic materials and articles intended to come into contact with foodstuffs. *Official Journal of the European Communities L* 220:18.
[21] European Commission (EC). 2005. Commission Directive 2005/79/EC of 18 November 2005 amending Directive 2002/72/EC relating to plastic materials and articles intended to come into contact with food. *Official Journal of the European Union L* 302:35.
[22] European Commission (EC). 2007. Commission Regulation 372/2007/EC of 2 April 2007 laying down transitional migration limits for plasticizers in gaskets in lids intended to come into contact with foods. *Official Journal of the European Union* L 92:9.
[23] European Commission (EC). 2008. Commission Regulation 597/2008 of 24 June 2008 amending regulation 372/2007/EC. *Official Journal of the European Union L* 164:12.
[24] Bueno-Ferrer C, Garrigós MC, Jiménez A. *Polym Deg Stabil*, (2010), DOI 10.1016/j.polymdegradstab.2010.01.027
[25] European Committee for Standardization (CEN). 2004. EN 13130-1, May 2004, Materials and articles in contact with foodstuffs. Plastics substances subject to limitation. Guide to test methods for the specific migration of substances from plastics to foods and food simulants and the determination of substances in plastics and the selection of conditions of exposure to food simulants.
[26] Goulas AE, Zygoura P, Karatapanis A, Georgantelis D, Kontominas MG. *Food Chem Toxicol.* 2007;45:585–591.

In: Biodegradable Polymers ...
Editors: A. Jimenez and G. E. Zaikov

ISBN 978-1-61209-520-2
© 2011 Nova Science Publishers, Inc.

Chapter 6

EVALUATION OF PARAMETERS ESSENTIAL FOR EFFICIENCY IN THE COMPOSTING PROCESS

J. Klein, M. Zeni, V. T. Cardoso, B. C. D. A. Zoppas, A. M. C. Grisa and R. N. Brandalise[*]

Universidade de Caxias do Sul, Rua Francisco Getulio Vargas, Bloco V – 205, CEP 95070-560 – Caxias do Sul/ RS, Brazil

ABSTRACT

The degradation of oxo-biodegradable polymeric materials after their use can be evaluated when they are exposed to different mediums, such as landfill, compostage; in direct contact with selected colonies of microorganisms, and others. The biological environment in which the polymers are generally disposed includes the presence of microorganisms. They use various sources of food and polymers may provide important substrates to obtain energy and produce new cells. This study proposes to look at the evolution of the degradation of organic matter by compostage, with PE films containing prooxidant additives (1 and 1,5 wt%), by evaluation of some parameters such as pH, humidity,

---

[*] E-mail: rnbranda@ucs.br

volatile solids, organic carbon, total nitrogen and temperature, as well as the fungi and metals characterization.

Keywords: Compostage; degradation; composting process; oxo-biodegradable

INTRODUCTION

Microorganisms (billions of them per gram of soil) are present in the environment. The biological environment in which the polymers are generally disposed includes the presence of microorganisms and they are essential for the polymeric biodegradation processes [1]. Microorganisms can use the carbon products to extract chemical energy for their life processes by:

1. Breaking the material (carbohydrates, carbon products) into small molecules by secreting enzymes or by environmental conditions (temperature, humidity, sunlight).
2. Transporting the small molecules inside the microorganisms cell.
3. Oxidizing the small molecules inside the cell to CO_2 and water while releasing energy that can be utilized by the microorganisms for its life processes in complex biochemical processes involving participation of three metabolically interrelated processes (tricarboxylic acid cycle, electron transport, and oxidative phosphorylation) [2].

Composting can be defined as the biological decomposition and stabilization of organic substrates under controlled conditions with transformation of biodegradable materials into a humus-like substance called compost, which can be beneficially applied to land [3]. Composting is a process controlled by those factors affecting microbial metabolism (i.e., temperature, moisture content, pH, nutrient content, etc.) [4].

High temperatures in aerobic decomposition during composting are not only caused by microorganisms but they are also an important factor determining their activity. Each microbial species can only grow within a certain temperature range, and most microorganisms are killed by temperatures too high. Mesophlic microorganisms are active up to 40-45 °C, while thermophlic organisms show optimum behaviour in the range 45-75°C. The temperature for maximum degradation rate in composting is normally near 55 °C, and the degradation rate is much lower at 70°C [5,6].

According to ASTM 6002 standard the mesophilic phase in composting occurs between 20 and 45 °C whereas the thermophilic phase occurs between 45 and 75 °C and it is normally associated with specific colonies of microorganisms that accomplish a high decomposition rate [7,8]. Decomposition of organic matter depends on the presence of water to support microbial activity (between 40 and 80%) [9]. During the first phase of the process the simple organic carbon compounds are easily mineralised and metabolised by microorganisms, producing CO_2, NH_3, H_2O, organic acids and heat.

The nutrient balance is very much dependent on the type of feed materials being processed. Carbon provides the preliminary energy source and the nitrogen amount determines the microbial population growth. Hence, maintaining the correct carbon/nitrogen (C/N) relationship is important to obtain a good quality compost. Bacteria, *actinomycetes*, and fungi require both carbon and nitrogen for growth. These microbes normally use 30 parts of carbon to 1 part of nitrogen [10].

During bioconversion of materials, carbon concentration will be reduced while nitrogen will increase in concentration, resulting in the reduction of the C/N ratio at the end of the composting process. This reduction can be attributed to the loss in total dry mass due to losses of C as CO_2. Ammonium-N (NH_4-N) and nitrate-N (NO_3-N) will also undergo some changes. NH_3 levels increase in the initial stages but declines towards the end [11].

pH value changes during composting due to changes in the chemical composition. In general, pH drops below neutral in the beginning due to the formation of organic acids and later rises above neutral because those acids are consumed and ammonium is produced [12].

In the last few years polyethylene (PE) films containing pro-oxidant additives (1 and 1.5 wt%) have been introduced in the market as a promising new material fulfiling biodegrability requirements in conjunction with the existing production and processing technologies. A significant portion of municipal and industrial waste consists of PE films used on a massive scale as wrapping material, a typical example being shopping bags.

Oxo-degradable PE is considered able to biodegrade in nature after 18 months [13,14]. Pro-oxidant additives represent a promising solution to the problem of environmental contamination by PE films. These substances can be transition metals complexes particularly based on Fe, Co [15] and Mn [16], and can increase the oxidation rate by air oxygen and cleavage of PE chains under the influence of light and/or heat. The polymer degradation, regardless of the mechanism, promotes changes in material properties, such as mechanicald and optical, and may cause cracks, erosion, discoloration and phases separation [17].

Some combined phenomena can accelerate degradation in synthetic polymers. They are photo-degradation, thermo-degradation, thermo-oxidation, and biological degradation This one is promoted by the action of bacteria, fungi and algae and may include contribution of some factors, such as humidity, polymer impurities, blending, presence of co-monomers and others [18].

This study proposes to check the evolution of organic matter degradation during composting of polyethylene films containing pro-oxidant additives without any initiation by light or heat. This work aims to contribute to the discussion on the influence of the degradation of PE films containing pro-oxidant additives in the composting, seeking thereby to infer the possibility of degradation not affecting the processing of organic compost.

EXPERIMENTAL

Materials

The compostage equipment was assembled at the University of Caxias do Sul (Brazil), and it was formed by a box with dimensions 100 x 120 x 60 cm^3 (Figure 1). Compostage soil was obtained by using a mixture of the organic fraction of urban solid waste and tree pruning in the proportions 75/25 % in mass, respectively. The blue PE films with dimensions 13 x 13 x 5x10^{-4} cm^3, were prepared in triplicate and placed in the composting medium. Samples were collected after 105 days and washed according to ASTM D6288-98 standard.

Testing

Some parameters for biodegradation of organic matter were monitored after the compostage in samples of the organic compost (~ 1 kg). Those parameters were pH (potentiometric method), humidity (gravimetry), volatile solids (gravimetry), organic carbon (Walkley-Black Method), total nitrogen (Kjeldahl titration), and temperature [19]. Metals (Mg, Mn, Cr, K, Na, Zn, Co,

Al, Ba, Ca, Cu, Fe and Pb) were monitored in an atomic absorption spectrometer (AAS), Mark Varian, spectrae 250 Plus.

During the 30-day period testing, fungi present in the compost and on the surface of polyethylene films containing pro-oxidant additives were recognised by their macro and microscopic characteristics based on the material incubated in Sabouraud dextrose agar, at 25 °C for 7 days. The methodology used in this study was based on the use of fungi culture collection reference [20,21]. The evaluation was carried out in an electron microscope (CARL ZEISS Axiostar).

RESULTS AND DISCUSSION

In a controlled aerobic biodegradation process it is possible to obtain organic carbon concentrations around 30 wt% at the beginning of test with further drop to around 10 wt% at the end of the process [22]. The main parameters corresponding to the organic matter in the compostage materials are shown in Table 1.

Figure 1. Compostage of blue PE prepared in University of Caxias do Sul/Brazil.

Table 1. Physico-chemical properties of samples tested

Parameters	prooxidant additives (%)	Time 0	(days) 10	20	30	40	60	105
	0%		5,39	8,25	8,52	8,9	8,65	Na
pH (25°C)	1%	4,77	7,12	8,32	8,36	8,24	7,89	Na
	1,5%		7,53	8	8,21	8,22	8,12	Na
	0%		84,71	81,01	79,97	80,71	82,12	81,5
Humidity (%)	1%	78,49	82,97	80,61	80,43	81,59	81,93	82
	1,5%		85,35	82,4	81,92	79,34	81,74	82,2
	0%		72,63	75,4	71,64	66,33	62,33	75
Volatile solids (%)	1%	78,18	72,9	75,96	68,55	69,1	69,05	68
	1,5%		74,11	74,51	78,68	69,25	59,81	77
	0%		84,6	58,42	52,67	50,21	49,09	49,9
Organic Carbon	1%	75,41	54,13	55,56	51,8	55,08	5,055	48,59
	1,5%		55,73	50,96	58,1	50,85	47,69	49,42
	0%		1,87	1,69	8,22	2,36	2,49	1,7
Total nitrogen (%)	1%	1,49	1,93	8,14	2,19	1,6	1,49	8,17
	1,5%		2,58	0,95	1,93	8,21	0,63	0,87
	0%		45,24	34,57	16,36	21,28	19,71	29,35
C/N	1%	50,61	28,05	18,01	28,65	34,43	38,93	18,75
	1,5%		21,6	58,64	29,07	16,84	75,7	58,8
	0%							
Temperature (°C)	1%	18(1)	36(3)	21(4)	16(2)	16(2)	14(2)	14(3)
	1,5%							

Na-not analyzed.

It was observed that temperature in the compostage medium raised from 18 ± 1 °C on the first day of the experiment to 41 ± 2 °C in just 24 hours, indicating the presence of thermopile microorganisms (45-75 °C) [23]. But from the tenth day of testing on the temperature decreased indicating the return to the mesophile microorganism's phase getting average temperatures for the rest of the experiment. It was also noted that the carbon/nitrogen (C/N) ratio varied from 50/1 to 16/1 during the 30 days of experiment, indicating that it the compost continues fully active after 30 days [22].

Figure 2 relates temperature C/N ratio with time, showing the limits of mesophilic and thermophilic phases. As the temperature decreased, the C / N ratio over time also decreased. The three typical phases of composting were observed during the process corroborating those reports by Haug [3]. A short initial mesophilic phase during approximately 2 days was observed and it was followed by a 2-weeks thermophilic phase, where temperature increased, and finally another mesophilic and maturation phase was detected after 25-30 days.

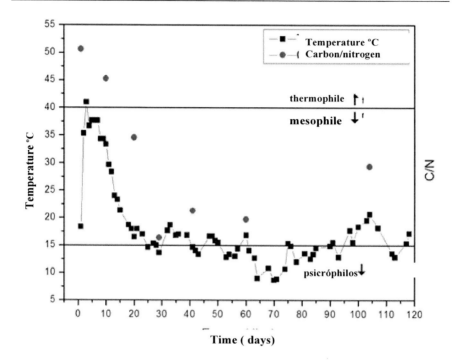

Figure 2. Relation between temperature and C/N ratio for composting.

It was also observed that pH of the organic matter degradation process varied from 4.8 to 8.5. According Rudnik [7], during the first stages of composting mesophilic bacteria and fungi degraded soluble and easily degradable compounds of the organic matter. Bacteria produced acids and consequently pH decreased. Proteins degradation leads to the release of ammonia and pH raised rapidly to 8-9.

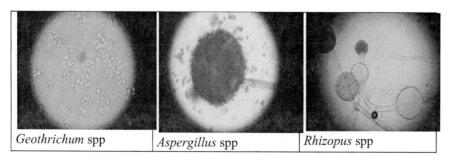

Figure 3. Fungi identified in the compost heap after 30 days of testing.

Humidity was varied from 78.5 to 80.0%, values considered high for the composting process, but this fact was attributed to the weather conditions during the experiment. However, humidity higher than 40% was considered adequate for efficient microbial activity. The volatile solids contents diminished from 78.2 to 71.6% indicating that the degradation process was taking place [24].

Different communities of microorganisms predominated during the succesive composting phases [22]. After 30 days, the macro and micro-morphology tests allowed the identification of yeast form cells, *hyaline hyphae,* and *arthroconidiae* characteristics of *Geotrichum spp* after 24 h for incubation. Fuurthermore it was possible to identify conidiophores and conidia characteristics of *Aspergillus spp, Penicillium spp* and *Fusarium spp*; sporangiophores and spores typical of *Rhizopus* spp. Fungi identified in the compost heap are shown in Figure 3.

The main results for characterization are indicated in Table 2.

All metals present in the pro-degradation additives were also found in the organic compounds (at 0 days) (Table 2). Cd, Pb, Co, Cr and Ni were detected neither in the additive nor in the organic compound. In addition Ba, Cr and K were identified in the pigment as well as in the additive and the organic compound. After 30 and 60 days of the experiment an increase in the concentration of Al, Ba, Ca, Fe, Mn, Na and Zn was detected. All these metals were found in high concentrations in the pro-degradation additive and pigment, and may contribute to the observed increase. Cu concentration remained constant throughout the testing period while Mg concentration showed a small reduction.

CONCLUSIONS

Some parameters were monitored to evaluate the efficiency of the organic matter degradation process in an experimental compostage device. The preparation of organic matter at a 75/25 ratio for wastes from urban residues was considered adequate to trigger the degradation process with C/N from 50/1 to 16/1 after 30 testing days. It was also found that the degradation process continues fully active at that time. The methodology to recognize fungal type microorganisms allowed the identification of *Geotrichum spp, Aspergillus niger, Rhizopus spp, Penicillium spp, Fusarium spp,* found in the first 30 testing days.

Table 2. Analysis of the metals of the organic compost of the pro-oxidant additive and of the pigment with 0, 30 and 60 days (standard deviation)

Elements	Organic Compost (mg/L)							Prooxidante additivie (mg/L)	Pigment (mg/L)	Detection limit (mg/L)
	0 days (white)	30 days			60 days					
		0%	1%	1,50%	0%	1%	1,50%			
Al	0,67	4,15	5,32	3,49	2,98	4,34	2,18	0,12(2)	100(2)	0,1
Ba	0,20	0,37	0,24	0,32	0,42	0,69	0,36	-	0,51(3)	0,1
Cd	-	Na	Na	Na	Na	Na	Na	-	-	0,02
Ca	133,75	196,35	88,55	90,91	149,9	438,85	115,35	0,7(1)	1348(10)	0,01
Pb	-	-	-	-	-	-	-	-	-	0,05
Co	-	-	-	-	-	-	-	-	-	0,02
Cu	0,13	0,10	0,11	0,12	0,13	0,13	0,13	0,02(1)	0,8(2)	0,01
Cr	-	-	-	-	-	-	-	-	0,050(0,00)	0,04
Fe	1,61	3,54	4,11	6,19	5,57	5,14	5,84	0,14(6)	2,55(5)	0,04
Mg	30,45	30,64	19,24	20,86	23,2	19,57	20,6	0,2(1)	157(22)	0,01
Mn	0,19	0,34	0,34	0,29	0,39	0,32	0,4	18(2)	0,2(2)	0,01
Ni	-	Na	Na	Na	Na	Na	Na	-	-	0,05
K	128,33	168,87	155,53	155,53	108,87	34,43	55,53	-	5,3(3)	0,01
Na	3,39	6,24	5,22	5,09	4,69	4,83	3,51	-	130(2)	0,01
Zn	0,25	0,29	0,29	0,3	0,35	0,33	0,39	0,013(6)	25,1(3)	0,01

Na – not analyzed.
Al – alumium ; Ba – barium; Cd-cadmium; Ca –calcium; Pb –lead; Co –cobalt; Cu – cooper; Cr – cromium; Fe – iron; Mg – magnesium.
Mn – manganese; Ni – nickel; K – potassium; Na –sodium; Zn –zinc.

From the environmental point of view the presence of Al, Ca, Cu, Fe, Mg, Mn and Zn was detected in the pro-degradation additive. These metals joined to those already present in the compound (Ba, K and Na). After 60 testing days the metals concentration increased with the exception of Cu which remained similar all through the test and Mg which decreased.

ACKNOWLEDGMENTS

The authors acknowledge the financial support from the Universidade de Caxias do Sul and CNPq (Brazil)

REFERENCES

[1] Miles DM, Scott G. *Polym. Degrad. Stabil.*, 2006;91:1581-1592.
[2] Narayan R. *BioPlastics Magazine*, www.bioplasticsmagazine.com, 4 (2009).

[3] Haug RT. *The Practical Handbook of Compost Engineering*. Lewis Publishers, Boca Raton, (1993).
[4] De Bertoldi M, Vallini G, Pera, A. *Technological aspects of composting including modelling and microbiology*. In: Gasser, JKR. (Ed.), Composting of Agricultural and Other Wastes. Elsevier Applied Science Publishers, London, pp. 27-40 (1985).
[5] Miller FC. *Composting as a process based on the control of ecologically selective factor* In: FBI Soil microbial ecology Meeting, Marcel Dekker, New York, 515-544 (1993).
[6] Ali Shah A, Hassan FH, Ahmed AS. *Biotechnology Advances* 2008;26:246-265.
[7] Rudnik E. *Compostable Polymer Materials*, Elsevier 2008, Amsterdam, 98 (2008).
[8] ASTM 6002 *Standard Guide for Assessing the Compostability of Environmentally Degradable Plastics*.
[9] Jeris JS, Regan RW, *Compost Sci.*, 1973;14:8-15.
[10] Pace MG, Miller BE, Farrel-Poe KL. *The Composting Process*. Extension, Utah State University. AG- WM 01 (1995).
[11] Liao PH, May AC, Chieng ST. *Bioresource Technology* 1995;54:159-163.
[12] Beck-Friis B, Smars S, Johnson H, Eklind Y, Kirchman H. *Composting Science & Utilization*, 2003;11:41-50.
[13] Koutny M, Lemaire J, Delort A. *Chemosphere*, 2006;64:1243-1252.
[14] Scott G, Willes DM. *Polym. Degrad. Stabil.*, 2006;91:1581-1592.
[15] Weiland M, Daro A, David C. *Polym. Degrad. Stabil.* 1995;48:275-289.
[16] Jakubowicz I. *Polym. Degrad. Stabil.*, 2003;80:39-43.
[17] Chiellini E, Corti A, Antone, SD, Bacuin, R. *Polym. Degrad. Stabil*; 2006;91:2739-2747.
[18] Gu JD, Ford T, Thorp K, Mitchell R. *International Biodeterioration & Biodegradation*, 1996;37:197-204.
[19] Shah A, Hassan F, Ahmed HS. *Biotechnology Advances* 2008;26:246-265.
[20] Barnett HI, Hubter BB. *Illustrated Genera of Imperfect Fungi*, MacMillan Publish Company, New York (1972).
[21] Lacaz CS, Porto E, Heins-Vaccari EM, Melo NT. *Guia para identificação: fungos, actinomicetos, algas de interesse médico*, Sarvier/Fapesp, São Paulo (1998).
[22] Siracura V. *Trends in Food Science & Technology*, 2008;19:634-643.

[23] Masaguer A, Moliner A, Hontoria CJ. *Bioresource Technology*, 2009;100:497-500.
[24] Mohee R, Unimar GD, Mudhoo A, Khadoo P. *Waste Management*, 2008;28:1625-1629.

In: Biodegradable Polymers ...
Editors: A. Jimenez and G. E. Zaikov

ISBN 978-1-61209-520-2
© 2011 Nova Science Publishers, Inc.

Chapter 7

CHARACTERIZATION OF PP FILMS WITH CARVACROL AND THYMOL AS ACTIVE ADDITIVES

M. Ramos[*], M. A. Peltzer and M. C. Garrigós
Analytical Chemistry, Nutrition & Food Sciences Department, University of Alicante, P.O. Box 99, 03080. Alicante, Spain

ABSTRACT

Polypropylene (PP) active films were prepared and characterized by incorporating thymol and carvacrol as active additives. Different concentrations were studied: 4, 6 and 8 wt% of carvacrol, thymol, and carvacrol and thymol (1:1) mixtures. A PP film without any active compound was also prepared as control. A complete characterization of all formulations was performed by testing different thermal, functional, structural and mechanical properties. TGA results showed that the addition of the studied additives did not affect the PP thermal stability. DSC confirmed the stabilization against thermo-oxidative degradation, with higher oxygen induction parameters obtained for materials with additives. Neat PP showed lower oxygen transmission rate (OTR) values than the active films. SEM images showed certain porosity for films with higher concentrations of thymol or carvacrol. Tensile tests showed

---
[*] E-mail: marina.ramos@ua.es

important differences between pure PP and formulations containing additives, in particular in elongation at yield and elastic modulus values, which could be due to a certain plasticization effect. The obtained results showed that these additives partially remained in the polymer matrix after processing and consequently they could be released during their shelf-life, acting as active additives in PP formulations.

Keywords: Polypropylene, active packaging, carvacrol, thymol, characterization

INTRODUCTION

Studies within the area of food packaging are experiencing a great development due to the consumer demands on safer and healthier products. Food active packaging systems are based on polymeric materials in which some additives with antimicrobial and/or antioxidant properties are added into the matrix with the aim of extending foodstuff shelf life [1-4]. Antimicrobial packaging is increasing the attention from the food and packaging industries due partly to increasing consumer demands for minimally processed and preservative-free products [5]. The incorporation of antimicrobial compounds into films results in increasing diffusion rates from the packaging material into the product, assisting the maintenance of high concentrations of the active ingredients where they are required. Furthermore, food-packaging films offer several advantages compared to direct addition of preservatives to food products since it is possible to control the release of active additives to the food [5-8].

Essential oils extracted from plants or spices are rich sources of biologically active compounds such as terpenoids and phenolic acids. It has been long recognized that some of the essential oils show antimicrobial properties [9]. Oregano essential oil is one of the most widely used and it is considered to be one of the most active plant extract against pathogens [10-12], being carvacrol and thymol its major components [13-14]. Carvacrol is present in high concentrations in oregano and thyme oils and exhibits a significant *in-vitro* antibacterial activity. This phenolic compound shows also antifungal, insecticidal, antitoxigenic and antiparasitic activities [15]. On the other hand, thymol is a phenolic monoterpene that has received considerable attention as an antimicrobial agent showing very high antifungal activity, being also an excellent food antioxidant [16-17].

The incorporation of these compounds to the packaging material leads to their gradual migration to foodstuff during storage and distribution operations. Consequently, antioxidant active additives incorporated into the polymer matrix can develop a double action; protecting the polymer during processing and also food during storage [18-19]. On the other hand, antimicrobial packaging can be effective in minimizing the superficial contamination of foodstuff, such as meat, fruits, vegetables, and so on [3; 17; 20-21].

The aim of the present study was the complete characterization of polypropylene (PP) films with carvacrol and thymol added at different concentrations to develop an antimicrobial/antioxidant active packaging system. The possible synergies between both compounds at the same concentration level was also evaluated.

EXPERIMENTAL

Materials

The polymer used in this work was PP ECOLEN HZ10K (Hellenic Petroleum, Greece) kindly supplied in pellets by Ashland Chemical Hispania (Barcelona, Spain) with melt flow index (MFI) 3.2 g/10 min (230 °C, 2.16 kg) and 0.9 g/cm^3 density. It was. Carvacrol (98 %) and thymol (99.5 %) were used as active additives and were purchased from Sigma-Aldrich (Madrid, Spain).

Sample Preparation

The different mixtures were obtained by blending the additives with the polymer into a Haake Polylab QC (ThermoFischer Scientific, Walham, USA) mixer at 190 °C for 6 min and a rotation speed of 50 rpm. The following formulations were prepared:

- PP without any active compound as control (PP0)
- PP containing 4, 6 and 8 wt% of carvacrol (PPC4, PPC6 and PPC8)
- PP containing 4, 6 and 8 wt% of thymol (PPT4, PPT6 and PPT8)
- PP containing 4, 6 and 8 wt% of carvacrol and thymol (1:1) (PPCT4, PPCT6 and PPCT8)

Films were obtained by compression-molding at 190 °C in a hot press (Carver Inc, Model 3850, USA). The material was kept between the plates at atmospheric pressure for 5 min until melting and then it was successively pressed under 2 MPa for 1 min, 3.5 MPa for 1 min, and finally 5 MPa for 5 min; in order to eliminate air bubbles. The average thickness of the films was around 200 μm measured by a Digimatic Micrometer Series 293 MDC-Lite (Mitutoyo, Japan) at five random positions around the film. The obtained films appearance was completely transparent and homogenous.

Characterization

Several analytical techniques were used in order to completely characterize the obtained films. Differential Scanning Calorimetry (DSC) was conducted using a TA DSC Q-2000 instrument (New Castle, DE, USA) under nitrogen atmosphere. Materials were exposed to the following thermal cycle: heating from 0 °C to 180 °C at 10 °C min^{-1}, followed by 3 min isothermal step and cooling at 10 °C min^{-1} to 0 °C, a further isothermal step for 3 min and heating to 180 °C at 10 °C min^{-1}. Melting temperatures (T_m) were determined from the second heating scan.

Oxygen induction time (OIT) was also determined by DSC at 200 °C in pure oxygen atmosphere. Initially, the material was heated under nitrogen (flow rate 50 mL min^{-1}) to the set temperature. After 5 min at 200 °C, the purge gas was switched to oxygen (50 mL min^{-1}). The heat flow was recorded in isothermal conditions up to the detection of the exothermic peak indicating the beginning of the oxidation reaction. OIT was determined by calculating time from switching the gas to the intersection between the baseline and the tangent of the exothermal oxidation signal.

Thermogravimetric Analysis (TGA) tests were performed in a TGA/SDTA 851 Mettler Toledo thermal analyzer (Schwarzenbach, Switzerland). Samples were heated from 30 °C to 700 °C at 10 °C min^{-1} under nitrogen atmosphere (flow rate 50 mL min^{-1}).

Oxygen transmission rate (OTR) is defined as the quantity of oxygen passing through the area unit of the parallel surface of a plastic film per time unit under pre-defined oxygen partial pressure, temperature, and relative humidity. In this case, pure oxygen was introduced into the upper half of the diffusion chamber while nitrogen was injected into the lower half, where an oxygen sensor was located. OTR was measured using a 8500 model Systech oxygen permeation analyzer (Metrotec S.A, Spain). 14-cm diameter circles for each formulation were clamped in the diffusion chamber at 25 °C an test was further started.

Cross sections of films were analyzed by scanning electronic microscopy (SEM) by using a JEOL JSM-840 instrument (Tokyo, Japan). An acceleration voltage of 10 KV was used. Films surfaces were previously coated with gold and images were registered at 500x magnification.

Tensile tests were carried out by using a 3340 Series Single Column System Instron Instrument, LR30K model (Fareham Hants, UK) equipped with a 2 kN load cell. Tests were performed with rectangular probes (dimensions: 100 x 10 mm^2), initial grip separation 60 mm and crosshead speed 25 mm min^{-1}. The resulting stress-strain curves were calculated according to ASTM D882-02 Standard procedure. Results were the average of five measurements.

RESULTS AND DISCUSSION

Thermal Properties

Thermal properties of samples were studied by DSC and four parameters were determined: crystallization temperature, T_c (°C); melting temperature, T_m (°C); crystallization heat, ΔH_c (J g^{-1}); and melting heat, ΔH_m (J g^{-1}). These parameters are summarized in Table 1. As can be observed, melting and crystallization temperatures as well as crystallization heat did not show appreciable differences between samples. Nevertheless, the melting heat of neat PP was higher than for the rest of formulations. This fact could indicate the higher crystallinity of the polymer when compared to formulations with additives. In this sense, crystallinity, χ (%), of samples was calculated by using the following equation:

$$\chi\ (\%) = (\Delta H_m\ /\ w.\Delta H_m°) \times 100\ \% \tag{1}$$

where the term $\Delta H_m°$ is a reference value representing the melting heat if the polymer were 100% crystalline; and w is the mass fraction of PP. $\Delta H_m°$ for PP has been reported to be 138 J g^{-1} [22]. Crystallinity results are also shown in Table 1, where a higher value for χ was obtained for the neat PP film (PP0). In conclusion, crystallinity of the material decreased with the addition of the studied antioxidants.

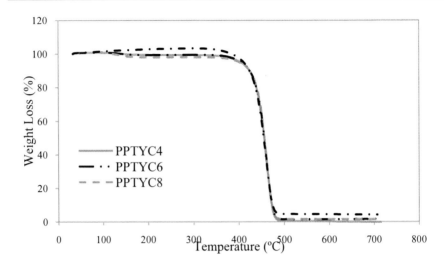

Figure 1. TGA curves obtained for PP0 and PP formulated with carvacrol and thymol (1:1).

Thermal and Thermo-oxidative Degradation

The effect of carvacrol and thymol on the thermal stability of films was studied by TGA under inert atmosphere. During thermal degradation, a single step was observed for PP0 sample. However, samples with carvacrol and thymol showed a first degradation step at lower temperatures (around 115 °C), which could be related to the volatilization of carvacrol and thymol, indicating the presence of both compounds remaining in the obtained films after processing. Figure 1 shows TGA curves for PP0 and PP with carvacrol and thymol (1:1), showing the additives loss at low temperatures. In this way, samples with higher concentrations of both compounds showed a more pronounced weight loss, as expected.

On the other hand, OIT analysis is the preferred test used in industry as the best indicator of polymer stability to oxidation. This parameter is commonly used to assess the oxidative stability of polyolefins [23]. The longer the OIT value the better the stabilization system used in the polymer is. Moreover, a higher OIT value indicates a higher amount of stabilizer remaining in the sample after processing.

OIT = $t_2 - t_1$, t_1 = 6.72 min.

Table 1. DSC results for PP formulations at different antioxidant concentrations

Sample	T_c (°C)	T_m (°C)	ΔH_c (J g^{-1})	ΔH_m (J g^{-1})	χ (%)
PP0	119	161	95.2	99.1	72
PPC4	119	161	91.4	49.6	37
PPC6	117	160	93.6	52.2	40
PPC8	118	161	89.2	48.1	38
PPT4	119	161	92.2	52.0	39
PPT6	117	160	86.5	47.3	36
PPT8	115	159	88.9	49.5	39
PPCT4	118	160	93.9	53.5	40
PPCT6	117	160	89.6	49.0	38
PPCT8	114	162	83.0	52.6	41

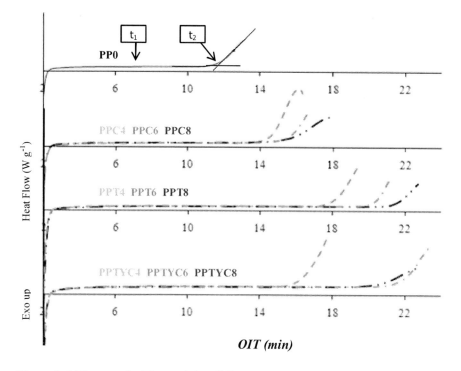

Figure 2. OIT curves for PP containing different natural additives at 200 °C.

Table 2. OTR results obtained for all formulations

Formulations	OTR.e (cm^3 mm m^{-2} day)		
Neat PP	82.2		
	4 wt%	6 wt%	8 wt%
PP + CARVACROL	104.7	146.9	159.6
PP + THYMOL	114.3	142.5	155.5
PP + (CARVACROL + THYMOL) (1:1)	103.1	112.6	112.2

(e: thickness, mm)

Figure 2 shows OIT curves obtained by DSC at 200 °C. As it can be observed, the studied additives clearly improve PP oxidative stability, since all of them increase the value of this parameter when compared to the result obtained for PP0. In general terms, OIT value rose with the increasing amount of additives. This effect was higher for those materials with thymol compared with those with carvacrol, indicating that thymol can be a best oxidative stabilizer for PP. In the case of those samples with equimolar amounts of carvacrol and thymol, they showed the highest stabilization efficiency, especially at 6 and 8 wt% concentrations, indicating a synergistic effect of both, carvacrol and thymol.

Oxygen Transmission Rate

Barrier properties to oxygen were studied by determining OTR.e values, where e is the film thickness in each material. Results are summarized in Table 2. As can be observed, PP0 films showed lower OTR.e values than those for films with carvacrol and thymol. A notable increase in OTR.e values was observed for PP with additives, in particular formulations with 8 wt% of carvacrol or thymol. On the other hand, a smaller increase was observed for formulations with both additives at equimolar concentration. The increase in OTR.e values could be due to the modification in the polymer structure when antioxidants are added, reducing the resistance of the film to oxygen transmission [24]. In addition, the decrease in the material crystallinity by the addtion of antioxidants (Table 1) could also lead to the increase in the gas transmission properties [25].

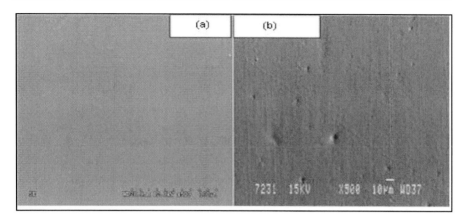

Figure 3. Scanning electron micrographs (500x) of surface films: PP0 (a) and PPCT8 (b).

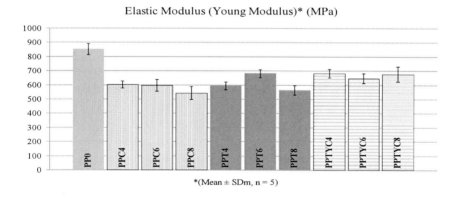

Figure 4. Elastic Modulus values for all formulations.

Morphological Characterization

The surface morphology of the obtained films was studied by SEM. As an example, Figure 3 shows SEM images for PP0 and PPCT8 films. As can be seen, an homogeneous surface was observed for neat PP film (Fig 3a). In the case of the film with additives, some porosity on the material surface was observed. This porosity could be due to the presence and possible evaporation of carvacrol and thymol from the polymer matrix. This effect was observed for all films with additives. The presence of these pores could be responsible for the higher oxygen transmission previously described.

Tensile Properties

Tensile tests were performed to study the effect of both additives, thymol and carvacrol, on the elastic modulus of all PP materials. Results are shown in Figure 4, where a significant decrease in elastic modulus can be observed for PP with additives compared to PP0 sample, being this effect more pronounced for PPC8 and PPT8 films. Apparently, this decrease in elastic modulus could be explained by the plasticizing effect of additives. Therefore, it could be said that the mechanical performance of the PP formulations with carvacrol and thymol was affected by the presence of these compounds.

CONCLUSIONS

Carvacrol and thymol were used at different concentrations up to 8wt% to prepare active films based on PP. Characterization of active films was performed using different analytical techniques in order to evaluate the effect of these additives in the polymer matrix and their stabilization performance. The addition of carvacrol and thymol did not affect significantly the thermal behaviour of PP but both additives modified oxygen barrier and mechanical properties of films. PP films with carvacrol and thymol showed an appreciable increase in OIT values, meaning that the polymer is well stabilized and, most important, that a certain amount of both compounds remained in the polymer matrix after processing, being able to play a roll as active additives for food packaging. It can be concluded that the addition of antioxidant and antimicrobial additives, such as carvacrol and thymol, to PP shows some potential to be used for improving product quality and safety aspects for food contact film applications.

ACKNOWLEDGMENTS

Authors would like to thank the Instituto Alicantino de Cultura Juan Gil Albert for financial support and Ashland Chemical Hispania for kindly supplying ECOLEN HZ10K PP (Hellenic Petroleum).

REFERENCES

[1] Del Nobile MA, Conte A, Buonocore GG, Incoronato AL, Massaro A, Panza O, *J. Food Eng.*, 2009;93:1-6.
[2] Álvarez MF, *Food Sci. Technol. Intern.*, 2000;6: 97-108.
[3] Appendini P, Hotchkiss, JH, *Innovative Food Sci. Emerging Technol.*, 2002;3 113-126.
[4] Vermeiren L, Devlieghere F, van Beest M, de Kruijf N, Debevere, J, *Trends Food Sci. Technol*, 1999;10: 77-86.
[5] Nerín C, López P, Sánchez C, Batlle R, *J. Agricultural Food Chem.*, 2007;55: 8814-8824.
[6] Kristo E, Biliaderis CG, Zampraka A, *Food Chem.*, 2007;101: 753-764.
[7] Suppakul P, Miltz J, Sonneveld K, Bigger SW, *J. Food Sci.*, 2003;68: 408-420.
[8] Peltzer M, Wagner J, Jiménez A, *Food Additives Contaminants*, 2009;26:938-946.
[9] Burt S, *Internat. J. Food Microbiol.*, 2004;94:223-253.
[10] Lambert RJW, Skandamis PN, Coote PJ, Nychas GJE, *J. Appl. Microbiol.*, 2001;91:453-462.
[11] López P, Sánchez C, Batlle R, Nerín C, *J. Agricultural Food Chem.*, 2007;55:4348-4356.
[12] López P, Sánchez C, Batlle R, Nerín C, *J. Agricultural Food Chem.*, 2005;53:6939-6946.
[13] Xu J, Zhou F, Ji BP, Pei RS, Xu N, *Letters in Applied Microbiol.*, 2008;47:174-179.
[14] Chizzola R, Michitsch H, Franz C, *J. Agricultural Food Chem.*, 2008;56:6897-6904.
[15] Veldhuizen EJA, Tjeerdsma-van Bokhoven JLM, Zweijtzer C, Burt SA, Haagsman HP, *J. Agricultural Food Chem.*, 2006;54:1874-1879.
[16] Youdim KA, Deans SG, *British J. Nutrition*, 2000;83:87-93.
[17] Sánchez-Garcia MD, Ocio MJ, Giménez E, Lagarón JM, *J. Plastic Film Sheeting*, 2008;24:239-251.
[18] Pérsico P, Ambrogi V, Carfagna C, Cerruti P, Ferrocino I, Mauriello G, *Polym. Eng. Sci.*, 2009;49:1447-1455.
[19] Peltzer M, Wagner J, Jiménez A, *J. Thermal Anal. Cal.* 2007;87:493-497.
[20] Lundbäck M, Hedenqvist MS, Mattozzi A, Gedde UW, *Polym. Degrad. Stabil.*, 2006;91:1571-1580.

[21] Nerín C, Tovar L, Salafranca J, Sánchez C, *J. Agricultural Food Chem.*, 2005;53:5270-5275.

[22] Joseph PV, Joseph K, Thomas S, Pillai CKS, Prasad VS, Groeninckx G, Sarkissova M, *Comp. Part A: Applied Sci. Manufacturing*, 2003;34:253-266.

[23] Cerruti P, Malinconico M, Rychly J, Matisova-Rychla L, Carfagna C, *Polym. Degrad. Stabil.* 2009;94:2095-2100.

[24] Sothornvit R, Krochta JM, *J. Agricultural Food Chem.*, 2000;48:3913-3916.

[25] Amstrong RB, *Tappi Place Conference*, 2002;285-311.

In: Biodegradable Polymers ...
Editors: A. Jimenez and G. E. Zaikov
ISBN 978-1-61209-520-2
© 2011 Nova Science Publishers, Inc.

Chapter 8

LIPASE CATALYZED SYNTHESIS OF BIOPOLYESTER AND RELATED CLAY-BASED NANOHYBRIDS

Hale Öztürk[1,2], Eric Pollet[1,], Anne Hébraud[1] and Luc Avérous[1,†]*

[1]LIPHT-ECPM, University of Strasbourg, 25 rue Becquerel, F-67087 Strasbourg Cedex 2, France
[2]Department of Chemical Engineering, Istanbul Technical University, Maslak 34469, Istanbul, Turkey

ABSTRACT

Biosynthetic pathway, like enzymatic ring opening polymerization (ROP) of lactones, attracts attention as a new trend of biodegradable polymer synthesis due to its non-toxicity, mild reaction requirement and recyclability of immobilized enzyme. Besides the enzyme-catalyzed synthesis of biopolyesters, key researches are conducted nowadays on the elaboration of biocomposites in combination with inorganic (nano)particles. The goal is to improve some of these polyesters properties for specific biomedical applications. In parallel, the use of

---
[*] E-mail: eric.pollet@unistra.fr
[†] E-mail: luc.averous@unistra.fr

clays as inorganic porous supports to immobilize enzymes has also been described. This chapter aims at presenting the use and development of original catalytic systems based on lipases which are efficient for polyester synthesis and allowing the preparation of hybrid materials based on clay nanoparticles grafted with such polyesters. For this, ε-caprolactone (ε-CL) polymerization catalyzed by Candida antarctica lipase B (CALB) was carried out in the presence of montmorillonite and sepiolite clays to obtain organic/inorganic nanohybrids through polymer chains grafting and growth from the hydroxyl groups of the clay. Both the free form and immobilized form of CALB have been tested as catalytic systems and their efficiency has been compared. The polymerization kinetics and resulting products were fully characterized with NMR, SEC, DSC and TGA analyses.

Keywords: Enzymatic polymerization; lipase; polycaprolactone; clay; nanohybrids

INTRODUCTION

Lipases are ubiquitous enzymes and have been found in most organisms from microbial, plant and animal kingdom [1]. Lipases are esterases which can hydrolyze triglycerides (or esters) to glycerols and fatty acids at the water-oil interface. In nature these enzymes take part in the degradation of food and fats by their hydrolytic capability. They also find application as valuable drugs against digestive disorders and diseases of the pancreas, as detergent additive for removal of fat stains and as catalysts for the manufacture of speciality chemicals and for organic synthesis [2].

Interestingly, in some cases, the lipase-catalyzed hydrolysis in water can be easily reversed in non-aqueous media into ester synthesis or transesterification [3]. This specific behaviour paved the way for the development of lipase-catalyzed polymerization reactions such as polyesters synthesis. Lipases catalyze the ring-opening polymerization (ROP) of lactones (small to large rings), cyclic diesters (lactides) and cyclic carbonates to produce aliphatic polyesters or polycarbonates [2]. The enzymatic polymerization can be regarded as an environment-friendly synthetic process for polymeric materials, providing a good example of "green polymer chemistry".

Indeed, some of the advantages of lipase-catalyzed polymerization are summarized hereafter. First, enzymes are derived from renewable resources, they are recyclable eco-friendly materials and can be easily separated from the synthesized polymers. They can be used in bulk, organic media and at various interfaces. Enzyme-catalyzed reactions take place under mild reaction conditions (temperature, pressure, pH, etc) with high enantio- and regio-selectivity. Polymers with well-defined structures can be formed by enzyme-catalyzed processes. In addition, lipases do not require the exclusion of water and air when used as catalysts for polyester synthesis. This is in contrary to the use of traditional organometallic catalysts where strict precautions must be taken to exclude air and water from the system. Moreover, only small (4–7 member) cyclic lactones having ring strains can be easily polymerized by organometallic catalysts whereas the polymerization of large and unstrained lactones (macrolides) is slow and only lead to low molecular weight products. On the contrary, lipases have shown the capability to polymerize both small lactones and macrolides under normal polymerization conditions [2] (Figure 1).

Finally, the organo-metallic catalysts used for the ROP of lactones or lactides are based on derivatives of metals such as Zn, Al, Sn or Ge, which may be very toxic [4]. Thus, the required complete removal of these metallic residues is an important issue when considering biomedical applications of these biopolyesters, since these toxic impurities may become concentrated within matrix remnants after polymer degradation [5]. Enzymes are natural catalysts and therefore better candidates for the elaboration of such biopolyesters for biomedical applications.

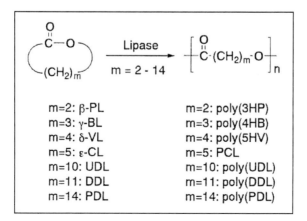

Figure 1. Lipase catalyzed ring-opening polymerization of lactones [6].

Thus far, the majority of enzymes studied for polyester synthesis have been from the lipase family. Certain lipases, stable in organic solvents, have shown a good ability to catalyze the esterification reactions for the synthesis of polyesters and polycarbonates. Among these, lipase B from Candida Antarctica (CALB) has proven to be a dominant catalyst for a wide range of these biotransformations [7,8]. CALB is a globular α/β type protein with a molecular weight of 33 kDa and an isoelectric point of 6.0. It is not as efficient as other lipases in hydrolyzing triglycerides; still, it is highly stereospecific towards both ester hydrolysis and synthesis. This phenomenon is probably related to the limited space available in its hydrophobic pocket [9].

Indeed, it was revealed that a unique structural feature is common to most lipases. An amphiphilic peptidic loop, called a lid or flap, is covering the active site of the enzyme in solution and preventing the access of the substrate. It was suggested that interfacial activation might be due to the presence of this lid. When a contact occurs with a lipid/water interface, this lid undergoes a conformational rearrangement which renders the active site accessible to the substrate [10]. On the contrary, the lipases from Pseudomonas glumae and Candida antarctica (type B), whose tertiary structure is known, both have an amphiphilic lid covering the active site but do not show interfacial activation [9]. After the lid is opened, a large hydrophobic surface is created to which the hydrophobic supersubstrate (oil drop) binds. It has been demonstrated that the active site is composed of a nucleophilic serine (Ser) residue activated by a hydrogen bond in relay with histidine (His) and aspartate (Asp) or glutamate (Ser105 - His224 - Asp187) [1,11].

Thus, the postulated mechanism for the lipase-catalyzed ring opening polymerization of lactones is shown in Figure 2 and can be described as follows. The carbonyl group of lactone is first subjected to the nucleophilic attack of lipase serine residues which are situated on the active center of the enzyme. This results in the ring-opening of a first monomer unit and the formation of an acyl-enzyme complex also called enzyme-activated monomer complex (EAM). Then water, which is present in small amount in the enzyme, acts as a nucleophile and reacts with the EAM intermediate, yielding oxyacid which is considered as the basic propagation species. In the propagation step, the terminal hydroxyl group of the growing polymer chains attacks the EAM to give a product that is extended by one additional repeating unit. The rate-determining step is the formation of the EAM intermediate [6,12,13].

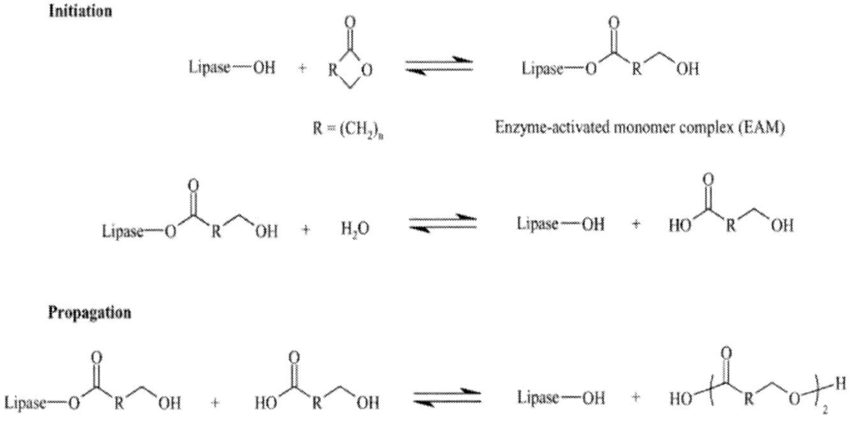

Initiation

R = (CH$_2$)$_n$ Enzyme-activated monomer complex (EAM)

Propagation

Figure 2. Postulated mechanism of lipase-catalyzed ring-opening polymerization of lactones [3].

Various process parameters of enzyme catalyzed reactions, such as operational and thermal stability and easier separation from the reaction mixture, can be modified by enzyme immobilization, [14]. Immobilization of lipases is often performed by adsorption through hydrophobic interactions between the enzyme and the support [15]. The enzyme binding depends on the nature of the surface and may be the result of ionic interactions, physical adsorption, hydrophobic bonding, Van der Waals attractive forces, or a combination of these interactions [16]. The enzyme immobilization can also be performed by covalent bonding through a single site or multiple sites anchoring and with or without the use of a spacer arm depending on the targeted flexibility of the system. Depending on the support and immobilization conditions, e.g. presence of additives, organic solvents or substrates, highly stable derivatives may be obtained [17].

Among the immobilized lipases, the most common and preferred system used is a physically immobilized form of CALB commercially available and known as Novozyme-435. This efficient and versatile catalyst is based on a porous acrylic resin as the enzyme carrier.

Alternative enzyme supports such as porous ceramic (silica) or other polymer-based supports for the development of immobilized lipase catalyst for efficient production of polymers are also investigated. For example, porous polypropylene was found to be a good support for Candida Antarctica in the polymerization of 15-pentadecanolide [6].

In parallel, the use of phyllosilicates as inorganic porous supports to immobilize enzymes has also been described. Interestingly, the use of such nanoclays as lipase carriers can bring numerous advantages, such as the high specific surface availability, the facility of water dispersion/recuperation, the high water uptake capacity and the excellent mechanical resistance of these materials [18].

Besides the enzyme-catalyzed synthesis of biopolyesters, important researches are conducted nowadays on the elaboration of composite materials in combination with inorganic (nano)particles to further improve some of these polyesters properties for specific biomedical applications. Till now, mainly hydroxyapatite, wollastonite and bioactive glass have been studied as the inorganic phase, mainly for the development of materials for implants or bone reconstruction. All studies show that these materials are very promising, the inorganic moiety bringing enhanced bioactivity and tailored degradation rate, but need further studies and improvement. Among the possible developments, the use of cheap and largely available clays is attentively considered.

Indeed, there has been a growing interest in the recent years for the elaboration of polymer-clay nanocomposites. Among the nanoclays, montmorillonite is by far the most studied for the elaboration of polyester/clay nanocomposite materials [19]. Montmorillonite is a crystalline 2:1 layered silicate mineral with general formula $(Na,Ca)_{0.33}(Al,Mg)_2(Si_4O_{10})(OH)_2 \cdot nH_2O$ and presenting a central alumina octahedral sheet sandwiched between two silica tetrahedral sheets (Figure 3) . The layer thickness is around 1 nm, and the lateral dimensions of these layers vary from 100 to 200 nm depending on the montmorillonite mining origin. Stacking of the layers leads to a regular van der Waals gap between the layers called the interlayer or gallery.

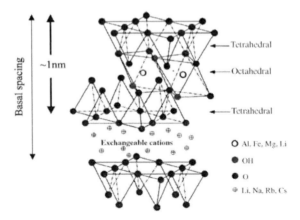

Figure 3. Structure of montmorillonite [20].

The isomorphic substitutions occuring within the clay platelets generate a negative charge naturally counterbalanced by inorganic cations (Na^+, Ca^{2+}...) located into the interlayer spacing giving a hydrophilic character to the native clay [20]. This interlayer spacing may vary from 1 to 3 nm depending on the hydration state of the inorganic cations.

As the forces that hold together the platelets stacks are relatively weak, the intercalation of small molecules between the layers is easy. In order to render these hydrophilic clays more organophilic, the hydrated cations of the gallery can be exchanged with cationic surfactants such as alkylammonium salts. When replacing the inorganic cations by organic cations such as bulky alkylammoniums, it usually results in a larger interlayer spacing. The modified clay (or organoclay) being organophilic, its surface energy is lowered and makes it more compatible with organic polymers. These polymers may be able to intercalate within the galleries, under appropriate experimental conditions.

Sepiolite is a non swelling, lightweight, porous clay with a large specific surface area and having particles of needle-like morphology. It is an uncommon clay because of its peculiar features and scarce occurrence in the nature. Sepiolite is a hydrated magnesium silicate with the following structural formula $Si_{12}Mg_8O_{30}(OH)_4(OH_2)_4 \cdot 8H_2O$. Sepiolite has a rather high surface area (around 400 $m^2\ g^{-1}$) and lower contact area between needles when it is compared to layered phyllosilicates. The fibers of sepiolite have an average length of 1-2 μm, a width of 0.01 μm and contain open channels of dimensions 3.6 x 10.6 Å running parallel to the fibers axis (Figure 4).

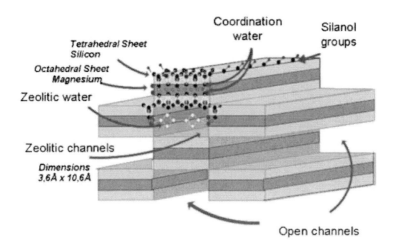

Figure 4. Structure of sepiolite [21].

These nanostructured tunnels are filled with zeolitic water bonded by hydrogen bonds at the external surface or within the channels under ambient conditions. The terminal Mg^{2+} cations located at the edges of the octahedral sheets provide their coordination with the two others structural water molecules. These molecules called as coordination water are in turn hydrogen-bonded to zeolitic water molecules [21,22]. Besides the peculiar needle-like morphology of the clay, one of the special features of sepiolite is the high density of silanol groups (-SiOH) available at the clay surface.

These two nanoclays are thus good potential candidates for the preparation of organic/inorganic hybrid materials based on biocompatible and biodegradable polyesters such as PCL. For this, nanohybrids of PCL grafted on montmorillonite and sepiolite clays surface were targeted by lipase-catalyzed ring-opening polymerization (ROP) of ε-caprolactone (CL)

EXPERIMENTAL PART

Materials

Cloisite® 30B is a commercial montmorillonite clay organo-modified by methyl bis(2-hydroxyethyl) (tallow alkyl) ammonium salts and supplied by Southern Clay Products (Texas, USA). The organo-modifier content of this clay is approximately 20 wt% as attested by thermogravimetric analysis. Each organo-modifier molecule bears two hydroxyl groups. Sepiolite (SEP) used in this study is a commercial clay (non organo-modified) supplied by Tolsa (Spain) under the tradename Pangel® S9 sodium.

Novozym 435 (NOV-435), immobilized form of CALB was purchased from Novozymes. The free form (not immobilized) of CALB, received as an aqueous solution, was kindly provided by Novozymes and was used after dialysis and lyophilization.

ε-caprolactone (ε-CL) (99%, Acros) was dried and stored over molecular sieves before polymerization. Anhydrous toluene was freshly distilled over sodium under nitrogen atmosphere.

Polymerization Reactions

All reactions were carried out in 2 mL of dry toluene at 70 °C. Predetermined amounts (typically 100 mg) of catalysts, NOV-435 or lyophilized CALB, were introduced into previously dried Schlenck tube under inert dry nitrogen atmosphere. The tube was immediately capped with a rubber septum. Toluene (2 mL) and ε-CL (1 mL) were transferred with a syringe through rubber septum caps. The reaction tube was then immersed in a heated oil bath and the polymerization reaction was allowed to proceed. An aliquot was withdrawn at specified time intervals to monitor the polymerization reaction progress. Reactions were terminated by dissolving the reaction mixture in chloroform and removing the catalyst by filtration. Chloroform in the filtrate was then stripped by rotary evaporation at 35 °C. The polymer in the resulting concentrated solution was precipitated in methanol. The polymer precipitate was separated by filtration and dried overnight at 30 °C under vacuum.

Nanohybrid Synthesis

Reactions were performed with the same procedure described above. The desired amount of sepiolite or montmorillonite (10 to 100 mg), dried overnight under vacuum at 80 °C or 60 °C respectively, was added in the reaction tube prior to the solvent and monomer addition. After the recovery of the product, the non-grafted PCL chains were removed from the nanohybrid by selective solubilization in chloroform using a Soxhlet apparatus for at least 48 h. The grafting efficiency, i.e., the percentage of grafted PCL was calculated from the inorganic content in the nanohybrid as determined by TGA.

Characterization

ε-CL polymerizations were monitored by proton nuclear magnetic resonance (^{1}H-NMR) to determine the monomer conversion and the number average degree of polymerization. ^{1}H-NMR spectra were recorded in CDCl$_3$ on a Bruker NMR spectrometer at 300 MHz.

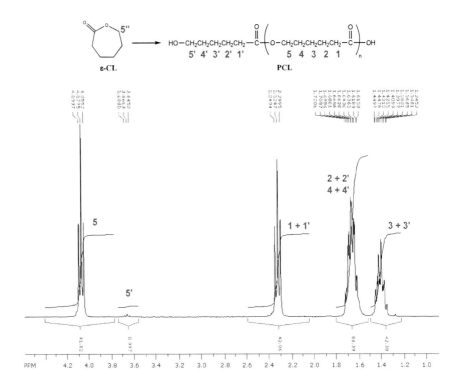

Figure 5a. Typical ^{1}H-NMR spectrum of PCL synthesized by lipase catalyzed ROP of ε-CL at 70 °C. Protons assignements and corresponding peaks integral areas are in the spectrum.

The molecular weight and polydispersity of PCL samples were determined by size exclusion chromatography (SEC) using a Shimadzu apparatus equipped with a RID-10A refractive index detector and an SPD-M10A UV detector. The columns used were PLGel Mixed-B 10 μm. The calibration was realized with PS standards ranging from 580 to 1.65 x 10^4 g mol^{-1}. Chloroform (puriss p.a. Riedel-de Haën) was used as the mobile phase and the analyses were carried out at 30 °C with a solvent flow rate of 1 ml min^{-1}.

Thermal characterization of the materials was carried out by thermogravimetric analysis (TGA) on 10-20 mg samples by heating from room temperature to 700 °C at 10 °C min^{-1} under He or air flow using a TGA Q5000IR apparatus from TA Instruments.

RESULTS AND DISCUSSION

Reference polymerization reactions of ε-CL were conducted with both free and immobilized lipase with the aim to determine the catalyst efficiency from the reaction kinetics and to characterize the obtained PCL. Aliquots of the reaction mixture were taken at specific time intervals, placed in NMR tubes and dissolved in CDCl$_3$ for ^{1}H-NMR analysis to monitor the polymerization reaction progress. A typical ^{1}H-NMR spectrum of PCL with the peaks assignments is shown in Figure 5a.

The chemical shifts in parts per million (ppm) are given relative to tetramethylsilane (TMS, 0.00 ppm) as the internal reference and they are the following 4.05 ppm (t,CH$_2$O), 3.65 ppm (t, CH$_2$OH, end group), 2.3 ppm (t, CH$_2$CO), 1.6-1.7 ppm (m, 2xCH$_2$), 1.35-1.45 ppm (m,CH$_2$). Figure 5b shows a zoomed view on the 3.5 to 4.4 ppm range which allows the determination of the monomer conversion and estimation of the degree of polymerization versus time.

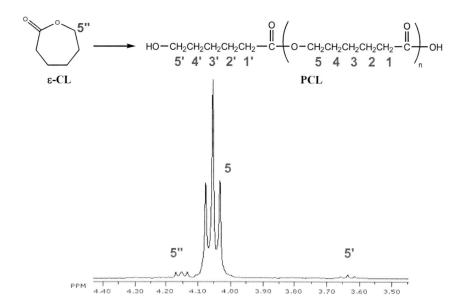

Figure 5b. Zoomed view of ^{1}H-NMR spectrum of a reaction mixture showing the methylene signals of monomer, polymer and chain-end.

Indeed, since the signal at 4.15 ppm (t, CH$_2$O) is assigned to the protons of ε-CL monomer, comparing its integral value (I$_{5''}$) to the integral of the corresponding signal for the polymer (I$_5$) allowing the calculation of the monomer conversion. In addition, the polymerization degree can be estimated from the ratio of integrals of the 5 and 5' signals at 4.05 and 3.65 ppm which are assigned to the CH$_2$O groups of the PCL main chain and the chain-end respectively.

These NMR analyses of aliquots from the reaction mixture allowed us to monitor the polymerization kinetics. Synthesis of PCL was first carried out with the immobilized form (NOV-435) and free form (CALB) of Candida Antarctica lipase, in absence of clay, as reference polymerization reactions. Then, polymerizations were performed in the presence of nanoclay and compared to the reference results obtained without clay.

In the case of ε-CL polymerization catalyzed by immobilized lipase NOV-435, the reaction proceeded fast with a monomer conversion greater than 95% within two hours (Figure 6a). Interestingly, the degree of polymerization and thus the number average molecular weight of the obtained PCL chains initially increased with monomer conversion and then reached a maximum value, after ca. 1 hour of polymerization time, before starting to decrease. Such a decrease in molecular weight observed at high monomer conversion and for long polymerization times is not surprising and it is more likely due to the well known inter- and/or intra-molecular transesterification reactions taking place in such conditions.

Regarding the polymerization of ε-CL catalyzed by "free" CALB previously dialyzed and lyophilized, a similar or slightly slower reaction rate was observed in comparison to the NOV-435 immobilized lipase. Besides, PCL chains of significantly lower molecular weights were recovered from the free CALB catalyzed polymerization when compared to those carried out with NOV-435 (Figure 7a). Nevertheless, a similar behaviour was observed with the slight decrease in molecular weights for long polymerization times.

It is worth pointing out that even if the added amount of NOV-435 catalyst is 100 mg, the estimated lipase content is only of ca. 10 mg (10% of the lipase immobilization) and this amount is much lower than for the free CALB catalyzed reaction. Thus, when considering the catalytic activity with the effective amount of enzyme, the NOV-435 immobilized system is more efficient than the non-immobilized counterpart.

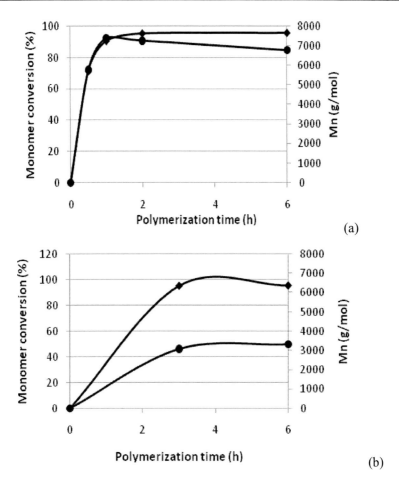

Figure 6. Time-evolution curves of ε-CL conversion (♦) and molecular weight (M_n) of obtained PCL (●), determined by ^{1}H-NMR analyses, for NOV-435 (100 mg) catalyzed polymerizations (a) in absence of clay and (b) in the presence of 10 mg of sepiolite.

Figures 6b and 7b show the characteristics of the free CALB and NOV-435 catalyzed polymerization reactions performed in presence of 10mg of sepiolite clay respectively.

With the addition of sepiolite to the reaction medium, one can see that the polymerization rate was not significantly affected since the monomer conversion was still higher than 95% after ca. 3 hours of polymerization.

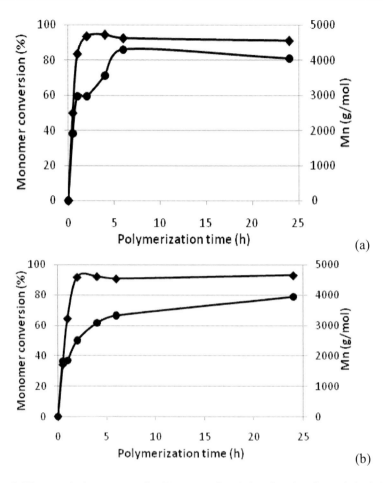

Figure 7. Time-evolution curves of ε-CL conversion (♦) and molecular weight (M_n) of obtained PCL (●), determined by ^{1}H-NMR analyses, for free CALB (50 mg) catalyzed polymerizations (a) in absence of clay and (b) in the presence of 10 mg of sepiolite.

Nevertheless, a decrease in the PCL molecular weight (M_n) was observed when sepiolite clay was added to the reaction medium. This decrease was more markedly observed for the NOV-435 immobilized lipase (Figure 6b). Lower PCL chain lengths were expected and can be explained by two main reasons. First, the addition of such hydrophilic clay may result in a higher water content in the reaction medium, which may lead to the initiation of more numerous PCL chains having thus lower average molecular weights. The second possible reason relies on the presence of numerous hydroxyl groups at the sepiolite surface acting as a co-initiator in the polymerization reaction.

Here again, this results in a higher number of growing chains and thus lower molecular weights for the obtained PCL. Interestingly, the participation of these hydroxyl groups in the polymerization reaction may also induce the grafting and growth of some polyester chains from the clay surface resulting in the formation of organic/inorganic nanohybrids.

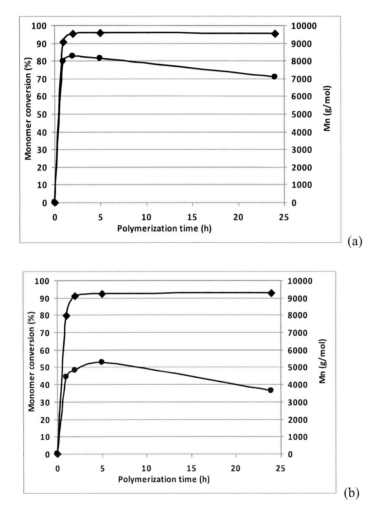

Figure 8. Time-evolution curves of ε-CL conversion (♦) and molecular weight (M_n) of obtained PCL (●) determined by ^{1}H-NMR analyses, for NOV-435 (100 mg) catalyzed polymerizations (a) in the presence of 10 mg of montmorillonite and (b) in the presence of 100 mg of montmorillonite.

For the elaboration of such organic/inorganic nanohybrids, the NOV-435 immobilized form of the lipase has been preferred and then it has been tested for the ε-CL polymerization in presence of montmorillonite clay which should have a lower content of surface hydroxyl groups. Figures 8a and 8b show the characteristics of the NOV-435 catalyzed polymerization reactions performed in presence of 10 mg and 100 mg of montmorillonite clay respectively.

As it can be seen, the addition of montmorillonite does not have a negative impact on the polymerization kinetics since monomer conversions higher than 90% were achieved within two hours. As expected, the lower content of hydroxyl groups available at the surface of montmorillonite platelets results in PCL molecular weights higher than those obtained in the presence of sepiolite. Interestingly, the PCL chains length obtained in the presence of 10 mg of montmorillonite and in absence of clay were very similar, attesting for the lack of negative impact from this nanoclay on the lipase polymerization activity. As already observed in the previous systems, the PCL molecular weight reached a maximum value and then showed a continuous decrease, for long polymerization times at high monomer conversion, likely due to transesterification reactions.

Figure 8b clearly highlights the impact of the clay content, and thus the amount of hydroxyl groups involved in the polymerization, on the PCL average molecular weight. Indeed, an increase in the montmorillonite content from 10 to 100 mg in the reaction medium resulted in a two fold decrease in the polymer chain length. This clearly attests for the effective participation of the clay hydroxyl groups on the PCL chain grafting and growth from the clay platelets.

In order to evidence such polymer grafting at the clay surface, the final product mixture was separated into a clay-rich fraction and a PCL-rich fraction by selective solubilization of the non-grafted PCL chains in chloroform using a Soxhlet apparatus. The solid clay-rich phase was then simply collected and dried whereas the non-grafted PCL was recovered from the solution by precipitation in methanol and further dried to obtain the PCL-rich phase. Both fractions were characterized by thermogravimetric analysis (TGA). Figures 9 and 10 show the weight loss curves of these PCL-rich and clay-rich phases, compared to the neat nanoclay, for the polymerization reactions carried out in the presence of respectively 100 mg of sepiolite (SEP) and 100 mg of montmorillonite (MMT).

Regarding the polymerization performed in the presence of sepiolite, the extracted PCL-rich fraction shows a multistep degradation process, with the main weight loss (more than 90%) occuring between 200 and 450 °C. The

PCL-rich phase recovered from the polymerization carried out with montmorillonite (Figure 10) displays a similar degradation behaviour except that it occurs at slightly higher temperatures probably by the higher molecular weight of the obtained PCL chains.

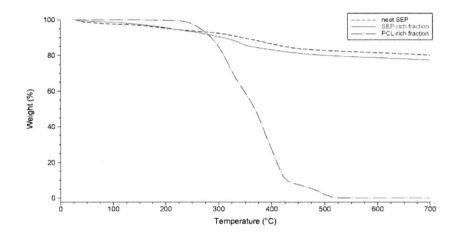

Figure 9. Weight loss curves of neat SEP, SEP-rich fraction and PCL-rich fraction obtained from TGA performed under air atmosphere.

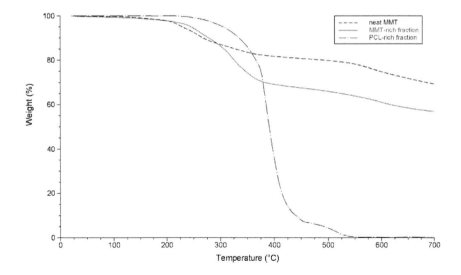

Figure 10. Weight loss curves of neat MMT, MMT-rich fraction and PCL-rich fraction obtained from TGA performed under air atmosphere.

As shown in Figure 9, neat sepiolite undergoes a multistep dehydration process when heated from 20 to 700 °C. The first step is due to the loss of hydration water (zeolitic water) which is physically bonded to sepiolite. This weight loss strongly depends on the atmospheric relative humidity. In the second step, sepiolite loses half of its coordinated water and then the remaining, more tightly bonded part, is lost in the third step occuring at higher temperatures. As it can be seen, the curve recorded for the SEP-rich fraction is very similar to the neat clay one but only differing by a slightly higher weight loss occuring between 250 and 400 °C. This temperature range corresponds to the PCL main degradation step and allows us to show that the SEP-rich fraction may contain approximately 3 wt% of grafted PCL chains.

As far as the polymerization in the presence of montmorillonite is concerned, the significant weight loss observed from 200 to 450 °C for the MMT-rich phase is likely due to grafted PCL chains degrading at these temperatures. According to the measured weight loss, we have estimated that the MMT-rich fraction contains approximately 12.5 wt% of grafted PCL chains. Such higher content of grafted chains can be explained by the lower content of water in such hydrophobic organo-modified montmorillonite compared to the hydrophilic sepiolite clay. In this case, one can assume that the abundant bonded water molecules can be involved in the enzymatic polymerization, hindering some hydroxyl groups of the sepiolite surface, and resulting in a high proportion of non-grafted PCL chains.

CONCLUSIONS

Two catalytic systems based on the immobilized and free-form of Candida Antarctica lipase B have been successfully tested for the ring-opening polymerization of ε-caprolactone. The commercially available immobilized lipase showed the highest efficiency and was selected for the elaboration of organic/inorganic nanohybrids based on PCL chains grafted at the surface of clay nanoparticles.

Determination of the polymerization rate showed that the addition of clay in the medium does not have negative impact on the kinetics of lipase-catalyzed polymerization of ε-CL. Nevertheless, the addition of clay results in a decrease in the PCL average molecular weights. Indeed, analyses evidenced a clear dependence between the decrease in PCL chains lengths and the increasing amount of hydroxyl groups available at the clay surface. The hydrophobic organo-modified montmorillonite, bearing less hydroxyl groups, leads to PCL chains with highest molecular weights.

The selective extraction procedures carried out on the reaction mixture and thermogravimetric analyses demonstrated that the hydroxyl groups at the clay surface are effectively involved in the polymerization reaction. This leads to the grafting and growth of PCL chains from the clay surface, resulting in the obtaining of PCL/clay nanohybrids, with the longer and higher amount of grafted chains being obtained with the montmorillonite clay.

Besides, on-going studies are carried out to physically immobilize CALB directly on sepiolite and montmorillonite clays aiming at the elaboration of efficient catalysts for the ROP of ε-CL and thus for the synthesis of well-defined organic/inorganic nanohybrids to be tested in biomedical applications.

REFERENCES

[1] Schmid R, Verger, R. *Angewandte Chemie International Edition,* 1998;37:1608-1633.
[2] Varma IK, Albertsson AC, Rajkhowa R, Srivastava, RK. *Progress Polym. Sci.,* 2005;30:949-981.
[3] Albertsson AC, Srivastava RK. *Advanced Drug Delivery Reviews,* 2008;60:1077-1093.
[4] Albertsson AC, Varma IK. *Biomacromolecules,* 2003;4:1466-1486.
[5] Schwach G, Coudane J, Engel R, Vert M. *Polym. Bull.,* 1996;37:771-776.
[6] Matsumura S. *Advances Polym. Sci.,* 2006;194:95-132.
[7] Mei Y, Kumar A, Gross R. *Macromolecules,* 2003;36:5530-5536.
[8] Hunsen M, Abul A, Xie W, Gross R. *Biomacromolecules,* 2008;9:518-522.
[9] Uppenberg J, Hansen MT, Patkar S, Jones TA. *Structure,* 1994;2:293-308.
[10] Brzozowski A, Derewenda U, Derewenda ZS, Dodson GG, Lawson DM, Turkenburg JP, Bjorkling F, Huge-Jensen BS, Patkar SA, Thim L. *Nature,* 1991;351:491-494.
[11] Jaeger KE, Dijkstra BW, Reetz MT, *Annual Review Microbiology,* 1999;53:315-351.
[12] Gross RA, Kalra B, Kumar A. *Applied Microbiology Biotechnology,* 2001;55:655-660.

[13] Kobayashi S, Takeya K, Suda S, Uyama H. *Macromol. Chem. Phys.* 1998;199:1729-1736.
[14] Gitlesen T, Bauer M, Adlercreutz PA. *Biochim. Biophys. Acta,* 1997;1345:188-196.
[15] Blanco RM, Terreros P, Pérez MF, Otero C, González G. *J. Mol. Catalysis B: Enzymatic,* 2004;30:83-93.
[16] Kennedy JF, White CA, Melo EHM. *Chimica Oggi,* 1998;5:21-29.
[17] Lafuente FR, Armisen P, Sabuquillo P, Fernández-Lorente G, Guisan JM. *Chem. Phys. Lipids,* 1998;93:185-197.
[18] Fuentes IE, Viseras CA, Ubiali D, Terreni M, Alcantara AR. *J. Mol. Catalysis B: Enzymatic,* 2001;11:657-663.
[19] Bordes P, Pollet E, Avérous L. *Progress Polym. Sci.* 2009;34:125-155.
[20] Alexandre M, Dubois P. *Mat. Sci. Eng.Reports,* 2000;28:1-63.
[21] Duquesne E, Moins S, Alexandre M, Dubois P. *Macromol. Chem. Phys.,* 2007;208:2542-2550.
[22] Tartaglione G, Tabuani D, Camino G. *Microporous Mesoporous Mat.,* 2008;107:161-168.

In: Biodegradable Polymers ...
Editors: A. Jimenez and G. E. Zaikov

ISBN 978-1-61209-520-2
© 2011 Nova Science Publishers, Inc.

Chapter 9

CHARACTERIZATION AND THERMAL STABILITY OF ALMONDS BY THE USE OF THERMAL ANALYSIS TECHNIQUES

Arantzazu Valdés-García[*], *Ana Beltrán-Sanahuja and M. Carmen Garrigós-Selva*

Analytical Chemistry, Nutrition & Food Sciences Department, University of Alicante, P.O. Box 99, 03080. Alicante, Spain

ABSTRACT

Nuts are subjected to thermal processes in the elaboration of manufactured products that can affect their thermal stability and lead to oxidation processes. In this paper, almond samples from three different cultivars (Spanish Guara; Marcona and Butte from USA) have been characterized by the use of Differential Scanning Calorimetry (DSC). Thermal stability of samples has also been evaluated by the use of Thermogravimetric Analysis (TGA). Fruit and oil samples have been studied at different scanning rates. Clear differences between Spanish and American almonds were observed by applying multivariate statistical analysis to DSC and TGA results, proving the suitability of the proposed methods for discrimination between different almond cultivars. Fruit and

---

[*] E-mail: arancha.valdes@ua.es

oil samples can be used for this purpose, although the use of fruit samples has the advantage of shorter sample preparation times.

Keywords: Thermal analysis, classification, characterization, almonds, multivariate analysis

INTRODUCTION

The quality and safety of processed foods depend, in part, on the packaging and the protection that it provides to foodstuff. Processing of fat foods at high temperatures leads to hydrogenation and esterification reactions, leading to the formation of *trans* fatty acids with negative modification of sensorial and nutritional properties, reduction of shelf-life affecting their overall quality and increasing the risk of cardiovascular diseases [1].

Characterization and thermal stability of food have been evaluated by using different methods and techniques, including thermal analysis. In this sense, Differential Scanning Calorimetry (DSC) has been used for characterization of different seed oils by obtaining their thermal profiles to avoid adulteration practises [2-3]. Oxidation kinetic profiles of different edible oils have also been reported [4-5] for the study of their oxidative stability, including the use of DSC [6]. Thermogravimetric Analysis (TGA) has also been used to study food samples behaviour subjected to different thermal treatments [7]. However, the use of these techniques for direct analysis of fruit samples [8-9] is currently limited in food industry being more usual the analysis of oil samples.

Nuts and specially almonds are very important food products from a nutritional point of view mainly due to their high content in numerous beneficial nutritive and bioactive compounds [10,11]. Almonds are grown as orchard crops, highly nutritious, but they have high cost. Despite their high fat content (46-76%) [12], almonds contain many valuable nutritional components [13]. Associated health benefits are attributed to their influence on the serum lipid profile and reduced cardiovascular risk in humans. These effects have been linked to their fatty acid composition (i.e. mainly mono and polyunsaturated fatty acids) and the presence of minor compounds with antioxidant activity (i.e. polyphenols and tocopherols) and cholesterol lowering effects of phytosterols [14-17].

In order to evaluate the efficiency of traditional packaging in the preservation of fat food, the characterization and study of their thermal degradation and conservation processes are necessary. Food-packaging active technologies with release of different antioxidants at controlled rates to reduce oxidation processes in a wide variety of foodstuff have been reported [18-21]. These active systems are able to interact with food or the associated headspace, positively changing the sensorial, nutritional and microbiological food properties, increasing shelf life and maintaining quality [22-24]. However, the initial study of food properties is not often considered in order to evaluate the packaging systems.

The aim of the present work was the thermal characterization and further classification of three different almond cultivars based on the use of thermal techniques (DSC and TGA). For this purpose, fruit samples and extracted almond oils were used and compared.

EXPERIMENTAL

Almond Samples

Twelve samples from three different almond cultivars were used in this work: four Marcona, four Guara and four Butte samples. Marcona and Guara cultivars were selected as they are representative of the cultivars grown in Spain. American Butte was chosen because it is one of the most widely grown cultivars in the world, mainly in California.

All samples were acquired unshelled from cultivars grown in the same crop year. Marcona and Guara cultivars were grown and collected from different Spanish geographical areas i.e. Xixona (Alicante), Pinoso (Alicante), Velez Rubio (Almería), Alcañiz (Teruel), Sella (Alicante), Tobarra (Albacete) and Barracas (Castellón). All of them were kindly supplied by "Colefruse S.A" (San Juan, Alicante, Spain). Butte samples were grown in California and were obtained from a Spanish importer (Almendras Llopis. San Vicente Del Raspeig, Alicante, Spain).

Sample Preparation

Shell of unshelled almonds was immediately removed by using a hammer. The obtained seeds were then stored at 7 ± 1 °C in order to be kept them fresh until use. Almond seeds were ground in a domestic electric grinder (Moulinex,

Barcelona, Spain) just before use or oil extraction. The seed fragments passing through a 1.5 mm sieve were stored in a desiccator.

For oil extraction, a previously developed analytical method was applied [17]. 5 g of ground almond seeds were extracted by using a commercial fat extractor (Selecta, Barcelona, Spain) with 40 mL of petroleum ether for 90 min. Temperature of the heating module was set to 135°C. The obtained oil was a mixture of twelve independent fat extractions and was dried under a nitrogen current and kept sealed in an amber vial in a freezer at -21 °C until required for analysis.

Differential Scanning Calorimetry (DSC)

Tests were performed in a TA Instruments Q2000 (New Castle, DE, USA) equipment. The purge gas was nitrogen at flow rate of 50 mL min^{-1}. Approximately 3-4 mg samples were weighed in hermetic aluminium pans (40μL) and were subjected to the following thermal program: sample loading at 50 °C, isothermal for 5 min, cooling the sample to -80 °C, isothermal for 5 min and further heating to 50 °C. Due to the strong dependence of thermograms on temperature-scanning rate, calorimetric tests were performed at 2, 5 and 10°C min^{-1}.

Thermogravimetric Analysis (TGA)

Dynamic TGA tests were carried out by using TGA/SDTA 851 Mettler Toledo (Schwarzenbach, Switzerland) equipment. Approximately 6-7 mg samples were weighed in alumina pans (70μL) and were heated from 30 °C to 700 °C at 10 and 5 °C min^{-1} under nitrogen atmosphere (flow rate 50 mL min^{-1}).

Statistical Analysis

Experimental data were processed with the aid of "SPSS statistical package Version 15.0" [25]. The presence of different categories within almond samples was investigated by using stepwise linear discriminant analysis (LDA) with the Wilks' lambda statistics for variable selection [26].

RESULTS AND DISCUSSION

DSC Analysis of Almond Oil and Fruit Samples

DSC analysis showed the presence of two energy transition events: an exothermic crystallization and endothermic melting. Heat flow curves obtained at different rates during cooling and heating for Guara Pinoso oil and fruit sample are shown in Figures 1 and 2, respectively. The same effect was observed for both samples: as cooling rate increased, temperature peak of exothermic transition moved to lower temperatures. At faster cooling rates, a certain amount of the fat crystallises in a less stable form at a lower crystallisation temperature, presumably because the temperature was lowered too fast for complete crystallisation of the stable form [27]. Moreover, as the heating rate increased, an increase in the transitions magnitude and peaks displacement to higher temperatures took place, but this displacement was lower compared with the cooling process.

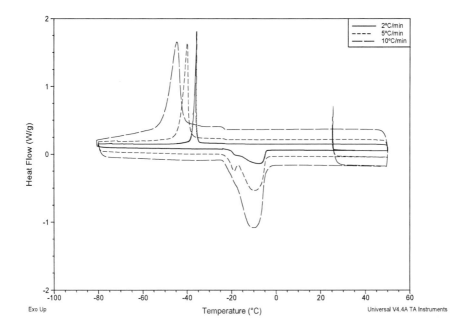

Figure 1. DSC curves obtained at different rates for Guara Pinoso oil sample.

The heating rate also affected the sharpness of endothermic peaks. A sharp melting peak is indicated by a narrow range of melting temperatures. In our studies, the endothermic processes were more pronounced in melting curves with increased heating rates. High scanning rates resulted in high melting points, owing to the poor oil thermal conductivity: at high scanning rates, there is no time enough for the heat to be transmitted from the heating elements of the DSC cell to the sample leading to an artificial shift of the melting curve [28].

Crystallization and melting parameters (temperatures and enthalpies) were calculated from the obtained DSC curves from each almond oil and fruit sample and they are shown in Table 1. Significant differences were obtained in samples from different almond cultivars, being higher when the analyses were carried out at 2 °C min^{-1} for oil samples and at 5 °C min^{-1} for fruit samples.

The main difference found for both types of samples was the value obtained for crystallization and melting enthalpies. In this sense, enthalpies for oil samples were twice higher than those obtained for fruit samples. This fact indicates that oil structure is more stable than in the case of the fruit. A similar behaviour was also observed for TGA results, as it will be further explained.

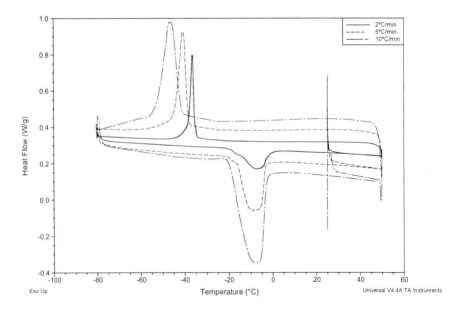

Figure 2. DSC curves obtained at different rates for Guara Pinoso fruit sample.

Table 1. Crystallization and melting parameters obtained from different cultivars almond oils and fruits at different rates

Sample Name	Analysis rate (°C min^{-1})	Almond oil T_c (°C)	ΔH_c (J/g)	T_m (°C)	ΔH_m (J/g)	Almond fruit T_c (°C)	ΔH_c (J/g)	T_m (°C)	ΔH_m (J/g)
Marcona	10	-46.9	57.1	-11.6	65.5	-50.2	20.5	-12.3	30.9
Butte	10	-50.7	52.6	-12.6	61.2	-52.1	19.0	-16.0	28.4
Guara	10	-45.1	56.7	-10.6	64.6	-48.4	17.3	-10.6	30.7
Marcona	5	-42.3	58.0	-10.9	65.3	-38.2	27.7	-9.9	28.1
Butte	5	-45.7	52.1	-12.6	59.5	-47.8	22.7	-11.6	26.1
Guara	5	-40.7	56.5	-9.9	63.7	-42.7	25.4	-8.4	31.5
Marcona	2	-37.7	59.8	-8.9	58.9	-38.6	30.2	-9.7	30.7
Butte	2	-40.8	58.9	-11.5	58.3	-43.8	28.9	-11.6	30.7
Guara	2	-35.9	64.0	-7.7	62.3	-43.2	26.8	-10.3	29.8

T_c (°C): crystallization temperature ΔH_c (J/g): crystallization enthalpy
T_m (°C): melting temperature ΔH_m (J/g): melting enthalpy

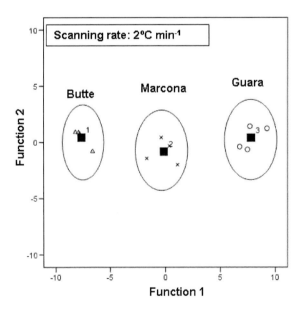

Figure 3. DSC mean scores of almond cultivars for the two discriminant functions obtained from the analysis of oil samples.

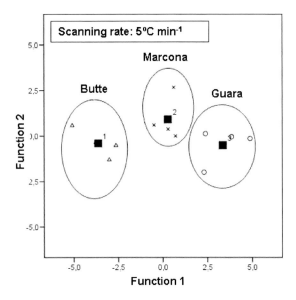

Figure 4. DSC mean scores of almond cultivars for the two discriminant functions obtained from the analysis of fruit samples.

In order to classify almond cultivars, a stepwise linear discriminant analysis using the Wilk's lambda statistics for variable selection was applied. This analysis led to two discriminant functions where 100 % of total variance was retained. For almond oils, melting temperature was the only variable not included in the analysis. On the other hand, for fruit samples enthalpies were not included. Figures 3 and 4 show the mean scores for the almond cultivars, which are projected on the reduced space of the two discriminant functions. By using the calculated discriminant functions, samples were correctly classified in all cases. It could be also observed that results obtained for fruit samples showed higher dispersion compared with those obtained for oil samples.

TGA Analysis of Almond Oil and Fruit Samples

Thermal stability of all samples was studied by using TGA (Figure 5). As can be seen, all curves showed a typical profile of sample weight loss occurring at different stages.

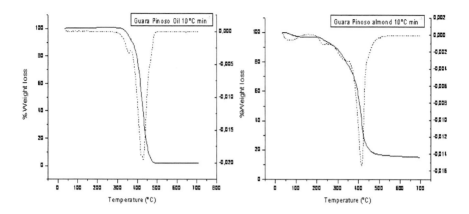

Figure 5. TGA and derivative curves of Guara Pinoso almond oil and fruit at 10° C min^{-1}.

For oil samples, the first step in TGAcurves represents the initial phase of the triglycerides degradation. In this stage, the oxidation of poly-unsaturated fatty acids takes place, starting at 289 °C and reaching its peak value at 352 °C. The second and third stages represent the monounsaturated and saturated fatty acids decomposition, respectively. The presence of antioxidants and the high quantity of saturated fatty acids make oils quite stable up to high temperatures.

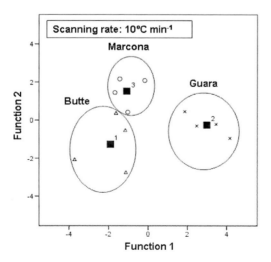

Figure 6. TGA mean scores of almond cultivars for the two discriminant functions obtained from the analysis of oil samples.

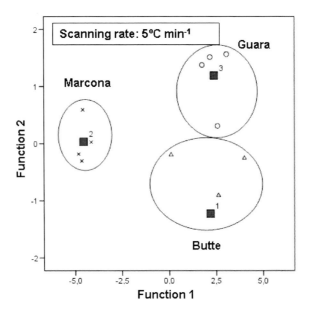

Figure 7. TGA mean scores of almond cultivars for the two discriminant functions obtained from the analysis of oil samples.

In the case of fruit samples, an initial degradation step was observed between 40 and 172 °C as the result of the volatiles and water losses. Three more stages were observed, corresponding to those appearing in oil samples. In particular the weight loss observed between 181 °C and 277 °C could be associated with oxidation of other compounds in samples, due to their complex matrix composition.

The peak of maximum degradation rate representing rupture of covalent bonds and structure changes with high mass loss (around 59 %) showed similar temperatures in both cases (about 400 °C). However, for the oil sample degradation began at higher temperatures than for fruit samples. This fact indicates that oil structure is more stable than in the case of fruit.

Significant differences were observed in oil samples from different almond cultivars, being higher when the analyses were carried out at 10 °C min^{-1} heating rate. Linear discriminant analysis led to two discriminant functions. The percentage of poly-unsaturated, mono-unsaturated and saturated fatty acids together with the initial degradation temperature of poly-unsaturated fatty acids were included in the analysis. Figure 6 shows the mean scores for almond cultivars, which are projected on the reduced space of the two discriminant functions. By using the calculated discriminant functions,

100 % of total variance was retained and samples were correctly classified in 91.7 % of the cases.

For fruit samples, significant differences were also observed from different almond cultivars, being higher when analyses were carried out at 5 °C min^{-1} heating rate. In this case, those variables included in the analysis were: weight loss, initial, final and maximum degradation temperatures of mono-unsaturated fatty acids. Figure 7 shows results of classification by using linear discriminant analysis. In all cases, 100 % of total variance was retained and the samples were correctly classified in all of the cases.

A comparison between TGA results with those obtained by DSC for samples classification, it can be observed that TGA showed higher dispersion for all samples (fruit and oil) and cultivars.

CONCLUSIONS

DSC and TGA have proved to be valuable techniques for almond oil and fruit authentication, by using multivariate data treatment. Both techniques have shown several advantages including small sample size, minimal sample preparation, short experimental times and no use of chemical reagents.

A complete DSC characterization has been performed by obtaining crystallization and melting parameters. Significant differences were observed in oil samples from different almond cultivars for the analyses carried out at 2° C min^{-1}. For fruit samples, the differences were highlighted at 5° C min^{-1}, indeed. Melting and crystallization temperatures were included in the analysis of fruit samples, but only melting temperature was considered in the case of oil samples. In conclusion, melting temperature can be used as a discriminant factor in order to classify almond samples according to the cultivar.

On the other hand, thermal degradation profiles have been obtained for the studied samples by TGA. The highest differences between almond cultivars in fruit and oil samples were observed for the analyses carried out at 5° C min^{-1} and 10° C min^{-1}, respectively. In this case, different variables were used depending on the sample.

DSC and TGA results indicated that oil presents a more stable structure than the almond fruit, although DSC has shown to be a more precise and sensitive technique compared with TGA. Finally, the use of fruit samples presents the additional advantage of a shorter sample preparation compared with the use of almond oil, which has to be extracted prior to analysis. In contrast, results obtained for oil samples have shown to be more reproducible than those obtained for fruits by DSC.

ACKNOWLEDGMENTS

Authors would like to thank "Almendras Llopis S.A" and "Colefruse S.A" for kindly providing almond samples.

REFERENCES

[1] Solís-Fuentes J, Camey Ortíz G, Hernández-Medel MR, Pérez-Mendoza F, Durán de Bazúa C. *Bioresource Technology*, 2010;101:799-803.
[2] Jiménez-Márquez A, Beltrán-Maza G, Aguilera-Herrera MP, Uceda-Ojeda M. *Grasas y aceites*, 2007;58:122-129.
[3] Miraliakbari H, Shahidi F. *J. Agric. Food Chem.*, 2008;56:4751-4759.
[4] Simon P, Colman L, Niklová I, Schmidt S. *J. Am. Oil Chem. Soc.*, 2000;77:639-642.
[5] Angiuli M, Bussolino GC, Ferrari C, Matteoli E, Righetti MC, Salvetti G, Tombari E. *Internat. J. Thermophysics*, 2009;30:1014-1024.
[6] Marikkar JMN, Lai OM, Ghazali HM, Che Man YB. *Food Chem.*, 2002;76:249-258.
[7] Diniz Z, Bora PS, Queiroga V, Oliveira JM, *Grasas y aceites*, 2008;59:160-165.
[8] Lodi A, Vodovotz Y, *Food Chem.*, 2008;110:554-561.
[9] Tironi VA, Tomás MC, Añón MC, *LWT - Food Sci. Techn.*, 2010;43:263-272.
[10] Mexis SF, Badeka AV, Kontominas MG, *Innovative Food Sci. Emerging Techn.*, 2009;10:580-589.
[11] Esfahlan AJ, Jamei R, Esfahlan RJ, *Food Chem.*, 2010;120:349-360.
[12] Mexis SF, Badeka AV, Chouliara E, Riganakos KA, Kontominas MG, *Innovative Food Sci. Emerging Techn.*, 2009;10:87-92.
[13] Jaceldo-Siegl V, Sabaté J, Rajaram S, Fraser GE, *British J. Nutrition*, 2004;92:533-540.
[14] Wijeratne SSK, Abou-Zaid MM, Shahidi F, *J. Agric. Food Chem.*, 2006;54:312-318.

[15] Sathe SK, Seeram NP, Kshirsagar HH, Heber D, Lapsley KA, *J. Food Sci.*, 2008;73:607-614.
[16] Garrido I, Monegas M, Gómez-Cordovés C, Bartolomé B, *J. Food Sci.* 2008;73:106-115.
[17] López-Ortiz CM, Prats-Moya S, Beltrán Sanahuja A, Maestre-Pérez SE, Grané-Teruel N, Martín-Carratalá ML, *J. Food Composition Anal.*, 2008;21:144-151.
[18] Granda-Restrepo DM, Soto-Valdez H, Peralta E, Troncoso-Rojas R, Vallejo-Córdoba B, Gámez-Meza N, Graciano-Verdugo AZ, *Food Research International*, 2009;42:1396-1402.
[19] Peltzer M, Wagner J, Jiménez A, *Food Additives Contaminants: Part A*, 2009;26:938-946.
[20] Mascheroni E, Guillard V, Nalin F, Mora L, Piergiovanni L, *J. Food Eng.*, Article in press (2010).
[21] Mayachiew P, Devahastin S, *Food Chem.*, 2010;118:594-601.
[22] Mexis SF, Kontominas MG, *Food Sci. Tech.* 2010;43:1-11.
[23] Mexis SF, Badeka AV, Riganakos KA, Kontominas MG, *Innovative Food Sci. Emerging Techn.*, 2010;11:177-186.
[24] Ercolini D, Ferrocino I, La Storia A, Mauriello G, Gigli S, Masi P, Villani F, *Food Microbiol.*, 2010;27:137-143.
[25] SPSS, release 15.0. 2007. SPSS Inc, Chicago.
[26] Tabachnik B, Fidell L, *Using Multivariate Statistics*, Harper Collins Publishers, 12th ed, 1992, 505-592.
[27] Che Man YB, Tan CP, *Phytochem. Anal.*, 2002;13:142-151.
[28] Tan CP, Che Man YB, *Phytochem. Anal.*, 2002;13:129-141.

In: Biodegradable Polymers ...
Editors: A. Jimenez and G. E. Zaikov

ISBN 978-1-61209-520-2
© 2011 Nova Science Publishers, Inc.

Chapter 10

CHITOSAN AS AN ANTIMICROBIAL AGENT FOR FOOTWEAR LEATHER COMPONENTS

M. C. Barros[1,*], I. P. Fernandes[1], V. Pinto[2,†], M. J. Ferreira[2], M. F. Barreiro[1] and J. S. Amaral[3,‡]

[1]LSRE, Bragança Polytechnic Institute, Campus de Santa Apolónia Ap 1134, 5301-857 Bragança, Portugal
[2]CTCP, Rua de Fundões - Devesa Velha, 3700-121 S. João da Madeira, Portugal
[3]REQUIMTE, Pharmacy Faculty, University of Porto, Rua Aníbal Cunha, 164, Porto 4099-030, Portugal and Bragança Polytechnic Institute, Campus de Santa Apolónia Ap 1134, 5301-857 Bragança, Portugal

ABSTRACT

Chitosan is being increasingly used in distinct areas such as pharmaceutical, biomedical, cosmetics, food processing and agriculture. Among the interesting biological activities that have been ascribed to chitosan, the antimicrobial activity is probably the one to generate the higher number of applications. Within this work the role of chitosan in diverse applications has been reviewed with particular emphasis for those

---

[*] E-mail: barros@ipb.pt, ipmf@ipb.pt and barreiro@ipb.pt
[†] E-mail: vera@ctcp.pt and MJoseF@ctcp.pt
[‡] E-mail: jamaral@ipb.pt

exploring its antimicrobial power. Furthermore, the mechanism to explain the antimicrobial activity of this emerging biopolymer is also discussed. The viability of using chitosan to effectively provide a functional coating for leather products was presented through an experimental case study. Results confirmed the potential of using this strategy to create antimicrobial leather products to be used, e.g., in the footwear industry.

Keywords: Chitosan; antimicrobial activity; functional coating, leather; footwear

CHITOSAN: STRUCTURE AND APLICATIONS

The use of biopolymers is currently attracting considerable attention for diverse applications due to their great potential, sometimes combining several properties like biodegradability, biocompatibility, non-toxicity and antimicrobial activity [1,2]. Examples of biopolymers are cellulose, starch, proteins, lignin and chitin. Biopolymers can be considered renewable since they are obtained from biomass which is available everywhere and in large amounts (by-products of agricultural, marine and forestry activities). The promotion of its use will contribute to create a more sustainable industry and society.

Chitosan is a cationic polysaccharide discovered in 1859 by Rouget and it is mainly produced by N-deacetylation of chitin, which is widely distributed in nature, as the structural component of the exo-skeletons of arthropods, including crustaceans and insects, in marine diatoms and algae, as well as in some fungal cell walls [3]. Structurally, chitin is a linear homo-polysaccharide consisting of N-acetyl-D-glucosamine repeated units, linked by β-(1-4) glycosidic bonds. On the contrary, chitosan is a linear hetero-polysaccharide comprising both D-glucosamine and N-acetyl-D-glucosamine linked by β-(1-4) glycosidic bonds. In fact, chitosan is a term that describes a heterogeneous group of polymers since they can present various deacetylation degrees (DD), different distribution of the acetamide groups within the polymer chain and various molecular weights. These structural differences will influence properties such as solubility and viscosity, and consequently its processability having in view industrial applications.

Like chitosan, chitin presents interesting characteristics like low toxicity and physiological inertness. Besides, chitin is being reported as presenting biological activity, like antimicrobial, antitumor and hemostatic activity and wound-healing acceleration [4,5]. Notwithstanding all of these desirable intrinsic properties, chitosan finds a wider range of applications, partly due to its polycationic structure that imparts solubility under acidic conditions. As a result, chitosan is being increasingly used in distinct areas such as pharmaceutical, biomedical, cosmetics, food processing and agriculture. A survey of some chitosan applications is shown in Table 1. Applications that arise from its antimicrobial activity will be presented in the next section.

ANTIMICROBIAL ACTIVITY OF CHITOSAN

All the interesting properties mentioned for chitosan have created an enormous attention, both by the scientific community and industry, to this natural and renewable resource. Among all, the antimicrobial activity is probably the property that triggered a higher number of scientific studies. The antimicrobial activity of different chitosans (obtained from different sources and tested under diverse conditions) has been reported by several authors [41,42]. The spectrum of activity includes both fungi (moulds and yeasts) and bacteria, being described to be more active against Gram-positive rather than Gram-negative bacteria [43]. In the last years, several mechanisms have been suggested to explain chitosan's antimicrobial activity. Nowadays, the action of chitosan is regarded to be complex, involving a series of events that ultimately can result in a bacteriostatic or bactericidal effect [43]. Although further research is still needed to clarify the underlying mechanisms, it is widely accepted that chitosan's antimicrobial activity is related to its polycationic structure [44]. It is believed that electrostatic interactions take place between the positive amino groups of chitosan and negatively-charged components of the cell envelope, such as teichoic and lipoteichoic acids in Gram-positive bacteria or phospholipids and lipo-polysaccharides of the outer-membrane in Gram-negative bacteria [43]. Those interactions would probably lead to changes in the cell wall integrity and permeability, resulting in modifications of trade flows between the inside and outside of the cell [4]. Besides interfering at cell wall level, some authors also suggested that chitosan can additionally interact with the cytoplasmic membrane, leading to a permeability change that would result in the release of some cellular components, such as enzymes and nucleotides [4], and even with genetic material such as DNA

[45]. The cytoplasmic membrane effects are generally considered to be a secondary event originated by interferences with the cell wall dynamics, such as the immobilization of lipoteichoic acids which ultimately would interfere in the lateral diffusion of proteins and fluidity of the cytoplasmic membrane [3,43]. Electron microscopy analysis supports this theory since observations are consistent with an impaired membrane function although the membrane remains intact, with no disruption or pore formation [3]. Regarding the possibility of interaction between chitosan and DNA, leading to the inhibition of m-RNA and protein synthesis, this would require the penetration of chitosan throughout the cell wall, which is composed of multilayers of cross-linked peptideoglycan, as well as through the cytoplasmic membrane. Considering that chitosan is a polymer that generally presents a very high molecular size, some authors consider this hypothesis rather unlikely to occur [43]. Moreover, Raafat and co-workers [3] did not find any evidence that chitosan could be broken down into smaller fragments by bacteria's enzymes, which might then pass into the cell.

Chitosan as an antimicrobial agent has found different applications in several industrial sectors such as food processing, textile, biomedical, pharmaceutical and cosmetics. Table 2 shows a summary of some relevant applications in these areas. Several factors can influence the extent of chitosan antimicrobial activity and it can be used in different forms (e.g. as hydrogels or as composites in biomedical applications, as films in the food industry or as microcapsules in cosmetics). All these factors and constraints must be considered when an application is envisaged.

The most reported factors that influence *in vitro* antimicrobial activity of chitosan are its molecular weight (M_w), deacetylation degree (DD), concentration, viscosity, pH, solubility, temperature and the tested microorganism [43,44]. Based on the available literature data, it is difficult to find a clear correlation between M_w and antimicrobial activity. Many studies have been performed on this subject; however it is difficult to stress out significant conclusions as variations can arise both from different DD and M_w distributions of chitosans, as well as from the tested strain, the growing stage of the culture (lag, log or stationary phase) and the microorganism concentration, among other factors.

Table 1. Summary of chitosan uses according to various fields of application

Application field	Functionality	Ref.
Biomedical/ Pharma-ceutical	Thermosensitive hydrogel for drug delivery	[6]
	Bioadhesive emulsions for drug delivery into or through body mucous membranes	[7]
	Lipid sequestration to reduce lipid digestion and absorption	[8]
	Biocomposites of chitosan/biomimetic calcium phosphate for orthopaedic, dental/craniofacial implants	[9]
	Epidermal substitute with permeation of polar and non-polar drugs	[10]
	Collagen/chitosan porous scaffold with improved biostability for skin tissue engineering	[11]
	Microcapsules/microspheres for drug delivery systems	[12], [13-17], [18]
	Hydrogel microbeads used for encapsulation of enzymes and cells or drug delivery	[19]
	Chitosan-coated PLGA nanoparticles for DNA/RNA delivery for genetic therapy	[20]
Food processing/ Nutrition	Chitosan microspheres for enzyme immobilization	[21]
	Chitosan microspheres for antioxidant delivery of olive leaf extract	[22]
	Chitosan based packaging to retard moisture, oxygen, aromas and solute transports	[23]
	Gelatin-chitosan films for improving shelf-life of cold-smoked sardine	[24]
	Stabilizer of tuna oil-in-water emulsions	[25]
Agriculture	Chitosan radiation-induced degradation products as growth promoter in agriculture fields	[26]
	Additive for coagulation/flocculation processes	[27]
	Transpiration reducer through foliar application	[28]
Cosmetics	Chitosan nanoparticles as carriers for retinol delivery	[29]
	Chitosan dispersions in glycolic acid suitable to formulate pharmaceutical/cosmetic products	[30]
	Irradiated PVAI membranes swelled with chitosan solution to be used as dermal equivalent	[31]
	Transparent xyloglucan–chitosan complex hydrogels for food and cosmetic applications	[32]
Other	Chitosan-based biodegradable insole	[33]
	Additive to improve dying processes (leather, textiles and wool)	[34], [35], [36-39]
	Chitosan-based polyols through oxypropilation	[40]

Table 2. Summary of chitosan uses as an antimicrobial agent according to various applications

Field of application	Functionality	Ref.
Biomedical/ Pharmaceutical	Development of anti-infective agents	[47]
	Lipid emulsion with antimicrobial activity for drug	[48]
	Gum supplementation for preventing dental caries	[49]
	Wound and/or burn dressing	[50-52]
	Biodegradable and biocompatible films for tissue engineering	[53]
	Microparticles for controlled release used in toothache and superficial skin wound treatment formulation	[54]
Food processing/ Nutrition	Delay of lipid oxidation and inhibition of bacterial growth in fresh pork sausages	[55]
	Improving shelf-life of food products	[24,56,57]
	Preventing agent to avoid bacteria beer spoilage	[58]
	Preservation/improved quality of strawberries	[59,60]
	Antimicrobial activity on fruit and vegetables	[61]
	Active packaging/films for food-borne pathogens inhibition	[62-65]
	Inhibition of surface spoilage bacteria in processed meats	[66]
	Improved antimicrobial activity and emulsifying properties	[67]
	Microbial growth inhibition in pasteurized milk and orange juice	[68]
Textile	Durable antimicrobial activity	[69]
	Antimicrobial activity at alkaline conditions	[70]
	Antimicrobial activity for wound healing	[2]
	Confer antimicrobial property to cellulose fibers	[71]
	Nanoparticles providing antibacterial and shrink-proof properties	[72]
Water treatment	Natural desinfectant	[73]

Lim et al [44] referred that, generally, the antimicrobial activity is thought to increase as the M_w of chitosan increases, but only to a certain value, from which the activity decreases. No and co-workers [41] tested the antibacterial activity of several chitosans (ranging from 28 to 1671 kDa) and chitosan oligomers (ranging from 1 to 22 kDa) against several Gram-positive and Gram-negative bacteria. In general, chitosans showed higher activity than chitosan oligomers, although it should be pointed out that the inhibitory effects differed with regard to the bacteria tested and the M_w of the chitosan. It also should be pointed out that in some cases (depending on the bacteria and M_w tested), weak or no antimicrobial activity was observed with some of the evaluated chitosans. In this study, the minimum inhibitory concentration (MIC) of chitosans ranged from 0.05% to > 0.1% depending on the bacteria and chitosan M_w [41]. The use of solutions with pH 5.5 prepared with chitosans ranging from 5 kDa to 165.7 kDa, with a DD of 88.76%, and microbial suspensions of *E. coli* and *S.aureus* (1.2×10^3 and 1.1×10^5 cells/mL, respectively) was reported by Zheng and Zhu [46]. They observed that for the Gram-negative bacteria (*E.coli*) the antimicrobial activity was enhanced as the M_w decreased, contrarily to the Gram-positive bacteria, for which the antimicrobial activity increased with increasing chitosan M_w. Regarding other important factors, it is generally accepted that the antimicrobial activity of chitosan is directly proportional to the DD, as well as to its concentration [44]. The increase of both factors implies a higher number of amino groups which can be protonated in acidic conditions, leading to an improved solubility and an increased chance of interactions between chitosan and negatively charged components in the microorganisms cell walls. Finally, pH should be taken into consideration as it strongly influences the deprotonation of the amino groups and the chitosan solubility [44].

THE RELEVANCE OF USING ANTIMICROBIAL LEATHER COMPONENTS IN FOOTWEAR

Consumers attitude towards hygiene and active lifestyle is creating new possibilities for the footwear industry, which in turn stimulates research and development. Microorganisms growth during use and storage of footwear can pose problems of material deterioration with associated unpleasant smell and even generate possible infections in susceptible individuals. Generally, footwear afford an excellent environment for the bacteria, yeasts and dermatophytic fungi growth since it presents high relative humidity, warmth and nutrients from feet sweat [74]. Additionally, leather itself and some tannery agents, such as oils and greases, provide a substrate where microorganisms can grow. In the foot, microtraumas caused by ingrown nails, abrasions and lacerations can allow microbial invasion through epidermis, resulting in skin infection [75]. This situation is especially important in diabetic patients since foot ulceration and amputation remains one of the most important causes of morbidity in diabetic people leading to frequent hospitalizations and considerable high social and economical costs [76].

Owing to its unique properties, chitosan can play an important role as an active agent for functional coatings. As referred previously, due to its antimicrobial properties, research concerning the use of chitosan as antimicrobial agent for industrial applications, such as in food, textile and leather industries is currently on-going. As footwear is prone to microbial attack, the development of functional antimicrobial coatings applied to footwear components that can directly contact with feet would be of great interest, both for the footwear industry (by reducing the possibility of material deterioration and quality loss) and from the consumer's point of view (by decreasing the possibility of skin infections and unpleasant odours).

CASE STUDY: CHITOSAN AS AN ANTIMICROBIAL AGENT FOR LEATHER COATINGS

In this work, the applicability of chitosan functional coatings to leather was tested with the purpose to develop new base materials to manufacture footwear components. To accomplish this objective, samples provided by the tanning industry at different tanning stages (wet-blue and lining leather) were treated with chitosan and studied for their antibacterial properties against 3 different bacteria (*Staphylococcus aureus, Pseudomonas aeruginosa* and *Escherichia coli*).

Leather Treatment with Chitosan

The treatment of samples with chitosan was carried out by following the next procedure: a chitosan solution (0.5%, w/v) was prepared by dissolving it (with a 70% deacetylation degree) in 2% (v/v) acetic acid with stirring for 1h at 50 °C. The obtained solution was then adjusted to pH 5 with NaOH and furthermore a leather sample (10 x 10 cm^2) was dipped in the solution for 30 min at room temperature. The sample was washed and dried in a stove at 60 °C. All samples were individually processed to obtain good homogeneity. A sample without any treatment was used as control. In order to evaluate the influence of the acid used in the chitosan solution, a second control was prepared by submitting a sample to the described procedure without adding chitosan.

Antimicrobial Activity Study

The antimicrobial activity was evaluated by using the Agar Diffusion Method based on the AATCC 147 test method [77] and both wet-blue and lining leather samples were tested. Three different bacteria were tested including Gram-positive (*Staphylococcus aureus*) and Gram-negative (*Pseudomonas aeruginosa* ATCC 27853 and *Escherichia coli* ATCC 10536) provided by the Microbiology Laboratory of Escola Superior Agrária de Bragança (Portugal) from its collection of isolated and identified strains. For each bacteria, an inoculum was prepared by transferring four morphological similar colonies from selective media to 10 mL of nutritive broth. Each broth was incubated at 35-37 °C (3 to 4 hours, depending on the bacteria) and the obtained suspension was adjusted to 0.5 McFarland standard density. Additionally, the absorbance at 625 nm was recorded for each inoculum. Bacterial solution were then transferred to the surface of a nutrient agar plate by making five consecutive parallel streaks (60 mm length, 10 mm distance from each other). The leather sample impregnated with chitosan (25 x 50 mm^2) was then placed transversely across the five streaks and the plate was incubated at 37 °C for 24 h. Similar tests using the same inoculum were performed for control samples.

Results and Discussion

Several tests are available to assess the antimicrobial activity of textile materials as such or after being subjected to an antimicrobial finishing treatment, including semi-quantitative and quantitative methodologies [78]. As no standardized microbiological procedures are available for the specific case of leather components, in this work the standard procedure to evaluate the antimicrobial activity of textile materials by the parallel streak method was followed [77]. This is a semi-quantitative method that is considered to be useful for estimating antimicrobial activity as the growth of the inoculum organism decreases from one streak to the next resulting in increasing degrees of sensitivity [77]. In this work, three different bacteria were tested: (*i*) *S. aureus*, which is a Gram-positive bacteria frequently found in the skin flora and that can be associated with skin and wounds infections [79] (*ii*) *P. aeruginosa*, which is a Gram-negative that can cause folliculites and dermatitis associated with hot-tubs and swimming pools contamination; it does not grow on dry skin, but it flourishes on moist skin [79] (*iii*) *E. coli*, which is the Gram-negative bacteria mostly used in this kind of studies. In order to evaluate the antimicrobial activity and to compare results, microbiological tests (using the control sample and the samples treated with and without chitosan) were simultaneously performed, i.e., using the same inoculum. This procedure guarantees that microorganisms were both at the same concentration and growing stage, thus allowing a reliable comparison of results. Figure 2 shows the results obtained for the wet-blue tests (control, sample treated without chitosan and sample treated with 0.5% chitosan) using the Gram-positive *S. aureus*.

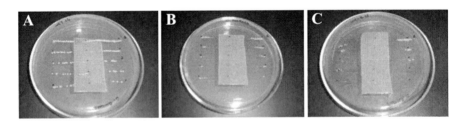

Figure 2. Paralell streak method using an *S. aureus* inoculum; A: wet-blue provided by the tanning industry (control); B: wet-blue submitted to treatment protocol but without no chitosan addition; C: wet-blue treated with 0.5% (w/v) chitosan solution.

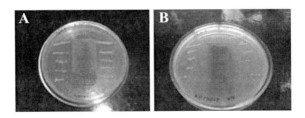

Figure 3. Paralell streak method using the *E. coli* inoculum; A: lining leather provided by the industry (control); B: lining leather treated with 0.5% (w/v) chitosan solution.

In the petri dish showing the control sample (Fig. 2A) it could be observed that microorganisms growth in the first streak (the one on the top) was more intense and that it decreased along the following ones. As referred, with this methodology the concentration of the inoculum gradually decreases from streak to streak, allowing to access to different sensitivity degrees. In the petri plate 2B, one can observe that the first streak showed a smaller inhibition zone than the others, which again can be explained by the higher microorganism concentration in the first streak and subsequent dilution in the following ones. As chitosan solution was prepared by using 2% acetic acid and this compound is known to present some antimicrobial activity [41], the test shown on Figure 2B was performed to estimate the contribution of acetic acid itself in the inhibition of the microorganisms growth. As it can be observed, the acetic acid contribution could not be discharged. This was expectable since acetic acid is a small molecule that can easily spread in the agar medium. Nevertheless, Figure 2C shows that the wet-blue treated with chitosan presented higher inhibition zones compared to the sample without chitosan,, with the last streaks showing almost complete inhibition demonstrating the effectiveness of chitosan as antimicrobial agent. Identical results were obtained in those plates inoculated with *E. coli* and *P. aeruginosa* showing that the studied chitosan coating presented antimicrobial activity both against Gram-positive and Gram-negative bacteria. The bacterial effectiveness of chitosan on Gram-positive and Gram-negative bacteria is still somewhat controversial, although recent works suggested that chitosan generally shows stronger effects on Gram-positive bacteria [41,42,44]. However, other authors claimed that hydrophilicity in Gram-negative bacteria is significantly higher than in Gram-positive bacteria, making them more sensitive to chitosan [42,43]. Goy and co-workers [42] referred that the charge density on the bacteria surface is a determinant factor to establish the amount of chitosan adsorbed, which in turn determines the impact on the cell manifested by greater changes in the structure and permeability of the cell envelope (cell wall and membrane). This suggests that the antimicrobial activity of chitosan can vary among microorganisms and even among different strains. Nevertheless, it should be stressed that variations among experimental procedures can also contribute to some dissimilarity among results. For example, the relevance of chitosan molecular weight was described by Eaton et al [80] when using atomic force microscopy (AFM) imaging to obtain high-resolution images of the effect of high M_w chitosan and chito-oligosaccharides (< 3 kDa) on the bacterial morphology of *E. coli* and *S. aureus*. Authors reported a stronger effect for the lower M_w chitosan in the case of the Gram-negative bacteria and the opposite in the case of the Gram-positive bacteria, thus evidencing the relevance of this factor to the antimicrobial effect on different types of bacteria. In our study, *S. aureus* seemed to be more affected by chitosan coating since it showed slightly higher inhibition zones compared to the other two studied bacteria. As referred, this could be possibly related to the cellular wall differences of Gram-positive and Gram-negative bacteria.

Figure 3 shows the results obtained with the lining leather test using the *E. coli* inoculum. In the control plate (Fig. 3A), it could be observed the inexistence of an inhibition zone in the first streak although weak inhibition is seen in the remaining streaks. During leather production, different chemical compounds are used, such as tanning agents and dyes. Possibly, some of those compounds can be retained in the lining leather and can spread in the agar during the test, inhibiting the microorganisms growth in the less concentrated streak and near the sample. Compared to control, the leather sample treated with chitosan presented significantly higher inhibition zones which demonstrates the antibacterial effectiveness of the chitosan film against *E. coli*. Identical results were obtained in the plates inoculated with *S. aureus* and *P. aeruginosa* showing that the studied chitosan film presented antibacterial activity both against Gram-positive and Gram-negative bacteria.

CONCLUSIONS

Both the wet-blue and lining leather treated with chitosan showed antimicrobial activity against the three tested bacteria, *E. coli*, *S. aureus* and *P. aeruginosa* thus presenting a high potential to be used as active coating material in prevention and control of bacterial growth in leather footwear components. Considering that several factors have been described to affect the extent of the antimicrobial activity of chitosan, namely molecular weight, deacetylation degree and pH among others, more testing should be performed by using chitosan with different characteristics in order to investigate possible increases in antibacterial activity. Results obtained, although preliminary, pointed out for the viability of using chitosan coatings to create antimicrobial leather products.

Acknowledgment

Financial support from COMPETE, QREN and EU (project QREN-ADI-1585-ADVANCEDSHOE) is acknowledged. ESA-IPB (Prof. Doutora Letícia Estevinho) kindly provided the *Staphylococcus aureus* used in this study.

References

[1] Takahashi T, Imai M, Suzuki I, Sawai, J. *Biochem. Eng. J.*, 2008;40:485-491.
[2] Shanmugasundara OL, *J. Textile Appare, Techn Management*, 2006;5:1-6.
[3] Raafat D, Bargen K, Haas A, Sahl H. *Applied Environ. Microbiol.*, 2008;74:3764-3773.
[4] Chung Y, Chen C. *Bioresource Technology*, 2008;99:2806-2814.
[5] Rinaudo M. *Progress Polym. Sci.*, 2006;31:603-632.
[6] Zhou HY, Chen XG, Kong M, Liu CS, Cha DS, Kennedy JF. *Carbohydrate Polym.*, 2008;73:265-273.
[7] Friedman D, Schwarz J, Amselem S. Bioadhesive emulsions preparations for enhanced drug delivery, US Patent 5993846 (1999).
[8] Helgason T, Gislason J, McClements DJ, Kristbergsson K Weiss J. *Food Hydrocolloids*, 2009;23:2243-2253.
[9] Chesnutt BM, Yuan Y, Brahmandam N, Yang Y, Ong JL, Haggard WO, Bumgardner, JD. *J. Biomedical Mat. Res. Part A*, 2007;10:333-353.
[10] Rana V, Babita K, Goyal D, Gorea R, Tiwary A. *Acta Pharmaceutica Sciencia*, 2004;54:287-299.
[11] Ma L, Gao C, Mao Z, Zhou J, Shen J, Hu X, Han, C. *Biomaterials*, 2003;24:4833-4841.
[12] Mao S, Shuai X, Unger F, Simon M, Bi D, Kissel T. *Intern. J. Pharmaceutics*, 2004;281:45-54.
[13] Xu JH, Li SW, Tostato C, Luo GS. *Biomedical Microdevices*, 2009;11:243-249.
[14] Ma L, Liu, C. *Colloids and Surfaces B: Biointerfaces*, 2009;75:448-453.
[15] Lu P, Wei W, Gong F, Zhang Y, Zhao H, Lei J, Wang L, Ma G. *Industrial Eng. Chem. Res.* 2009;48:8819-8828.
[16] Brunel F, Véron L, Ladavière C, David L, Domard A, Delair T. *Langmuir*, 2009;25:8935-8943.

[17] Pancholi K, Ahras N, Stride E, Edirisinghe M. *J. Mat. Sci.: Materials in Medicine*, 2009;20:917-923.
[18] Wei W, Ma G, Wang L, Wu J, Su Z. *Acta Biomaterialia*, 2009;6:205-209.
[19] Mark D, Haeberle S, Zengerle R, Ducree J, Vladisavljević GT. *J. Colloid Interface Sci.*, 2009;336:634-641.
[20] Nafee N, Taetz S, Schneider M, Schaefer UF, Lehr C. *Nanomedicine: Nanotechnology, Biology and Medicine*, 2007;3:173-183.
[21] Biró E, Németh AS, Tóth J, Sisak C, Gyenis J. *Chem. Eng. Processing*, 2009;48:771-779.
[22] Kosaraju SL, D`ath L, Lawrence A. *Carbohydrate Polymers*, 2006;64:163-167.
[23] Pranoto Y, Rakshit SK, Salokhe VM. *LWT - Food Sci. Technol.*, 2005;38:859-865.
[24] Gómez-Estaca J, Montero P, Giménez B, Gómez-Guillén, MC. *Food Chemistry*, 2007;105:511-520.
[25] Klinksorn U, Namatsila Y. *Food Hydrocolloids*, 2009;23:1374-1380.
[26] El-Sawy NM, Abd El-Rehim HA, Elbarbary AM, Hegazy, EA. *Carbohydrate Polymers*, 2010;79:555-562.
[27] Renault F, Sancey B, Badot PM, Crini G. *Eur. Polym. J.*, 2009;45:1337-1348.
[28] Bittelli M. Flury M, Campbell GS, Nichols EJ. *Agricultural and Forest Meteorology*, 2001;107:167-175.
[29] Kim D, Jeong Y, Choi C, Roh S, Kang S, Jang M, Nah J. *Intern. J. Pharmaceutics*, 2006;319:130-138.
[30] Anchisi C, Maccioni AM, Meloni MC. *Il Farmaco*, 2004;59:557-561.
[31] Rodas ACD, Ohnuki T, Mathor MB, Lugao AB. *Nuclear Instruments and Methods in Physics Research Section B: Beam Interactions with Materials and Atoms*, 2005;236:536-539.
[32] Simi CK, Abraham TE. *Food Hydrocolloids*, 2010;24:72-80.
[33] Catinari M. *Biodegradable insole*, EP 1676493 A1 (2005).
[34] Haroun AA. *Dyes and Pigments*, 2005;67:215-221.
[35] Burkinshaw SM, Jarvis AN. *Dyes and Pigments*, 1996;31:35-52.
[36] Houshyar S, Amirshahi SH. *Iranian Polymer Journal*, 2002;11:295-301.
[37] Dev VRG, Venugopal J, Deepika G, Ramakrishna S. *Carbohydrate Polymers*, 2009;75:646-650.
[38] Jocic D, Vílchez S, Topalovic T, Navarro A, Jovancic P, Julià MR, Erra P. *Carbohydrate Polymers*, 2005;60:51-59.
[39] Pascual E, Julià MR. *J. Biotechnol.*, 2001;89:289-296.

[40] Fernandes S, Freire CSR, Pascoal-Neto C, Gandini A. *Green Chemistry*, 2008;10:93-97.
[41] No HK, Park NY, Lee SH, Meyers, S.P. *Intern. J. Food Microbiology*, 2002;74:65-79.
[42] Goy RC, Britto D, Assis OBG. *Polímeros: Ciência e Tecnologia*, 2009;19:241-247.
[43] Raafat D, Sahl H. *Microbial Biotechnology*, 2009;2:186-201.
[44] Lim S, Hudson SM. *J. Macromol. Sci.- Polym. Rev.*, 2003;C43:223-235.
[45] Rabea EI, Badawy MET, Stevens CV, Smagghe G, Steurbaut, W. *Biomacromolecules*, 2003;4:1457-1465.
[46] Zheng L, Zhu J. *Carbohydrate Polymers*, 2003;54:527-530.
[47] Park Y, Kim M, Park S, Cheong H, Jang M, Nah J, Hahm K. *J. Microbiol. Biotechnol.*, 2008;18:1729-1734.
[48] Jumma M, Furkert FH, Müller BW. *Eur. J. Pharmaceutical Biopharmaceutics*, 2002;53:115-123.
[49] Hayashi Y, Ohara N, Ganno T, Yamaguchi K, Ishizaki T, Nakamura T, Sato M. *Archives of Oral Biology*, 2007;52:290-294.
[50] Vicentini DS, Smania A, Laranjeira MCM. *Mat. Sci. Eng. C*, In Press (Available online February 2009).
[51] Campos M, Cordi L, Durán N, Mei L. *Macromol. Symposia*, 2006;245:515-518.
[52] Wittaya-Areekul S, Prahsarn C. *Intern. J. Pharmaceutics*, 2006;313:123-128.
[53] Freier T, Koh HS, Kazazian K, Shoichet MS. *Biomaterials*, 2005;26:5872-5878.
[54] Aelenei N, Popa MI, Novac O, Lisa G, Balaita L. *J. Mat. Sci.: Materials in Medicine*, 2009;20:1095-1102.
[55] Georgantelis D, Ambrosiadis I, Katikou P, Blekas G, Georgakis SA. *Meat Science*, 2007;76:172-181.
[56] Alteri C, Scrocco C, Sinigaglia M, Nobile MA. *J. American Dairy Science Association*, 2005;88:2683-2688.
[57] Casariego A, Souza BWS, Vicente AA, Teixeira JA, Cruz L, Díaz, R. *Food Hydrocolloids*, 2008;22:1452-1459.
[58] Gil G, del Mónaco S, Cerrutti P, Galvagno M. *Biotechnology Letters*, 2004;26:569-574.
[59] Campaniello D, Bevilacqua A, Sinigaglia M, Corbo MR. *Food Microbiology*, 2008;25:992-1000.
[60] Vargas M, Albors A, Chiralt A, Gónzalez-Martínez C. *Postharvest Biology and Technology*, 2006;41:164-171.

[61] Devlieghere F, Vermeulen A, Debevere J. *Food Microbiology*, 2004;21:703-714.
[62] Zivanovic S, Chi S, Draughon DF. *J. Food Sci.* 2005;70:45-51..
[63] Zivanovic S, Li J, Davidson M, Kit K. *Biomacromolecules*, 2007;8:1505-1510.
[64] Fernández-Saiz P, Lagarón JM, Ocio MJ. *Food Hydrocolloids*, 2009;23:913-921.
[65] Fernández-Saiz P, Lagarón JM, Hernandez-Muñoz P, Ocio MJ. *Intern. J. Food Microbiology*, 2008;124:13-20.
[66] Ouatarra B, Simard RE, Piette G, Bégin A, Holley RA. *Intern. J. Food Microbiology*, 2000;62:139-148.
[67] Song Y, Babiker EE, Usui M, Saito A, Kato, A., *Food Research International*, 2002;35:459-466.
[68] Lee CH, Park HJ, Lee DS. *J. Food Eng.* 2004;65:527-531.
[69] Payne SA. Durable antimicrobial leather, US 2006/0014810 A1 (2006).
[70] Lim S, Hudson SM. *Carbohydrate Research*, 2004;339:313-319.
[71] Alonso D, Gimeno M, Olayo R, Vásquez-Torres H, Sepúlveda-Sánchez JD, Shirai K. *Carbohydrate Polymers*, 2009;77:536-543.
[72] Yang H, Wang W, Huang K, Hon M. *Carbohydrate Polymers*, 2010;79:176-179.
[73] Chung Y, Wang H, Chen Y, Li S. *Bioresource Technology*, 2003;88:179-184.
[74] Orlita A. *International Biodeterioration & Biodegradation*, 2004;53:157-163.
[75] Jennings MB, Alfieri D, Kosinski M, Weinberg JM. *The Foot*, 1999;9:68-72.
[76] Global Lower Extremity Amputation Study Group, *J. Surgery*, 2000;87:328-337.
[77] American Association of Textile Chemists and Colorists, *Antibacterial Activity Assessment of Textile Materials: Parallel Streak Method* (1998).
[78] Silva RM, Pinto VV, Freitas F, Ferreira MJ. *Multifunctional Barriers for Flexible Structure, Textile, Leather and Paper*, Chapter 13, Springer Berlin Heidelberg, 229-268 (2007).
[79] Qarah S, Cunha A, Dua P, Lessnau KD. *eMedicine*, URL: http://emedicine.medscape.com/article/226748-overview, (up-dated 2009).
[80] Eaton P, Fernandes JC, Pereira E, Pintado ME, Malcata FX. *Ultramicroscopy*, 2008;108:1128-1134.

In: Biodegradable Polymers ... ISBN 978-1-61209-520-2
Editors: A. Jimenez and G. E. Zaikov © 2011 Nova Science Publishers, Inc.

Chapter 11

CHARACTERIZATION OF LIGNOCELLULOSIC MATERIALS BY MORPHOLOGICAL AND THERMAL TECHNIQUES

M. I. Rico[1], M. C. Garrigós[2,*], F. Parres[a] and J. López[1]

[1]Mechanical and Materials Engineering Department, Polytechnic University of Valencia, 03801, Alcoy, Alicante, Spain
[2]Analytical Chemistry, Nutrition & Food Sciences Department, University of Alicante, P.O. Box 99, 03080, Alicante, Spain

ABSTRACT

The search for new materials is nowadays being focused on the use of biodegradable polymers such as lignocellulosic by-products. In this paper, different biomass products as nut shells and wood have been characterized in order to study their morphological and thermal behaviour. Scanning electron microscopy (SEM) has been used for the surface morphological study of different nut shells and wood. Structures of both samples have been compared in terms of possible expected mechanical properties. Sequential pyrolysis-gas chromatography-mass spectrometry (Py-GC-MS) was used for the thermal characterization of almond shells, where some characteristic volatile compounds were obtained with a decrease as sequential cycles were applied. The obtained

---

[*] Corresponding author. Email: mcgarrigos@ua.es

results provided useful information on the thermal degradation of cellulose and lignin obtained from almond shells.

Keywords: Wood, nuts, characterization, lignin, cellulose

INTRODUCTION

With the depletion of fossil fuel and environmental concerns, the utilization of biomass resources has achieved increasing worldwide interest [1]. Agricultural waste materials as nut shells and wood are a potentially important energy source [2]. Lignocellulosic by-products from pruning, debris, nut shells and seeds are abundant in food and timber industries residuals and therefore they have easy access and low prices. The use of this source of biomass through energy recovery processes would help to solve the disposal problem [3,4].

Another important application of biomass is the use as fillers for inclusion within polymer matrices [5]. Fillers could have different morphologies, normally being solid and finely divided, and they are added to many plastic formulations to modify their properties or to increase density. Lignocellulosic materials belong to the group of organic fillers. In recent years they have acquired great importance due to its biodegradability nature, being used in the automotive industry, residential products, and many other applications with growing markets, including polymer blending and composites applications [6-10].

In previous studies it was reported that the degradation of all biomass can be considered as the overlapping of three main components: cellulose, hemicellulose and lignin [11]. Therefore, studies on the characteristics of these components are important for a better understanding of the biomass degradation. In this sense, biomass pyrolysis has received special attention for studying its thermal degradation [12,13]. On the other hand, scanning electron microscopy (SEM) is a well-known technique for the study of the surface morphology of different materials, being also used for the characterization of biomass materials [14,15]. Chemical and morphological characterization of different biomass materials has been carried out, including some nut shells [16,17] and wood [18-20] by the use of the mentioned analytical techniques. However, the majority of these studies have been focused on materials derivatives.

Pyrolysis is basically a polymeric structure cracking process, to convert lignocellulosic materials into volatiles and char [21]. Temperature and heating rate are those process parameters with the largest influence on the pyrolysis products [22]. It is widely accepted that the primary pyrolysis of biomass materials generally takes place in the 200-400 °C range, resulting in bulk product volatilization, as well as the formation of a solid char residue. Once the temperature increases above 400 °C, products continue to evolve slowly, as the char residue undergoes further chemical and physical transformations [23].

The aim of the present work was the characterization of different lignocellulosic materials by using different analytical techniques in order to study their morphological and thermal behaviour. SEM will be used for the morphological study of different nut shells and woods, in order to compare the structures and mechanical properties of both types of natural samples. On the other hand, Py-GC-MS will be used for the study of the thermal degradation process of almond shells and mainly the volatile compounds generated and its relationship with cellulose/lignin compounds.

EXPERIMENTAL

Reagents and Materials

Cellulose and sulphonated lignin were purchased from Sigma-Aldrich (Madrid, Spain). Two types of lignocellulosic materials were used in this work. The wood species used were pine, oak, beech and sapelli in form of sawdust, all obtained from "Agloma S.L." (Alcoy, Alicante, Spain). Residuals in form of ground shells from different nuts (apricot, almond, hazelnut, peach and walnut) were supplied by "Jesol" (Onteniente, Valencia, Spain).

Material Characterization

The surface morphology of samples was studied by using an Olympus SZX7 stereomicroscope (Tokyo, Japan) and a JEOL JSM-6300 scanning electron microscope (Jeol USA Inc., Peabody, USA).
Pyrolysis products of lignocellulosic materials were studied from almond shells. A Pyroprobe® 1000 CDS pyrolyzer (Analytical, Inc., Oxford, PA)

coupled to an Agilent 6890N gas chromatograph 5973N quadrupole mass spectrometer (Palo Alto, CA, USA), operating in electronic impact (EI) ionization mode (70 eV) was used. Samples (around 1 mg) were pyrolyzed at 450 °C for 10 s and repeatedly subjected to sequential pyrolysis under the same conditions until 4 cycles were completed. A HP-5MS capillary column (30 m x 0.25 mm i.d. x 0.25 µm film thickness) (Agilent Techn., Santa Clara, CA, USA) and a split-splitless injector were used (split ratio 1:50). Helium was used as the carrier gas with 1 mL min^{-1} flow rate. Temperatures for injector and detector were 300 °C. The column temperature was programmed from 40 °C (hold 5 min) to 300 °C at 5 °C min^{-1} heating rate (hold 15 min). Identification of detected compounds was performed in full scan mode (40-650 m/z), by matching their mass spectra against NIST mass spectral library.

RESULTS AND DISCUSSION

Morphological Study

The morphology analysis of lignocellulosic materials is very useful for the determination of material properties. For this purpose, a preliminary analysis of the surface of each shell sample, where external, internal and cross section surfaces were studied, was carried out by using a stereomicroscope (magnification x4). It was observed that the external surface of the studied materials was completely different from the internal or cross section zones. Different structures were also observed for the studied nuts. It has to be pointed out that external surfaces are exposed to environmental factors that are dependent on the tree family, climate and area. On the other hand, all the internal surfaces showed a completely flat and brilliant structure, due to the permanent contact with seeds, not being affected by any environmental factor. For these reasons, it was decided to study the cross section of shells, since it is the only surface to give actual information about the internal structure of the material without being affected by external factors.

After this preliminary study, the observation of the surface morphology of shells cross sections was performed by using scanning electron microscopy (SEM). Figure 1 shows micrographs obtained for cross sections of all the studied shells.

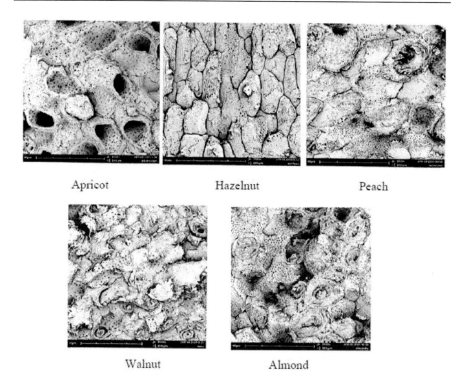

Figure 1. Scanning electron micrographs of nut shell surfaces (1500x).

As observed from Figure 1, all surfaces showed some irregularities joined to a remarkable roughness, being the apricot shell the more porous structure with a higher number of irregularities and cavities formed from concentric layers. The peach and walnut shells showed structures very similar to those of apricot shell, but with lower number of cavities. In contrast, hazelnut shell showed a granular and quite homogeneous structure with high degree of orientation, and no cavities. Finally, the almond shell also showed many cavities with a higher number of concentric layers. This type of structure makes shells with a high aspect ratio very important for their use as fillers in polymer matrices.

The obtained micrographs were analyzed by using the image processing software Scandium M, obtaining data shown in Table 1. As can be observed, the material with highest cavities area was the peach shell; whereas apricot presented the highest grains area. On the other hand, the almond shell showed the highest thickness of concentric layers, in contrast with hazelnut shell where this type of structure was not observed.

Table 1. Data obtained from the analysis of micrographs of different nuts

Nut sample	Cavities/grains area (μm^2)	Concentric layers thickness (μm)	Cavities perimeter (μm)
Apricot	674 ± 235	5 ± 1	106 ± 18
Walnut	406 ± 165	8 ± 2	86 ± 19
Peach	1362 ± 243	5 ± 2	162 ± 14
Hazelnut	794 ± 178	-	124 ± 10
Almond	711 ± 382	14 ± 6	102 ± 29

Surface morphology of sections For wood samples was observed by using SEM (Figure 2). As can be observed, longitudinal section of all materials showed strongly oriented fibrillar structures with porosities and cavities, but in a lower extent than those observed for shell structures. The absence of high levels of porosity makes wood a material with good mechanical properties. In contrast, when studying the cross sections of wood, a well oriented and with no porosities honeycomb structure was observed (Figure 2C).

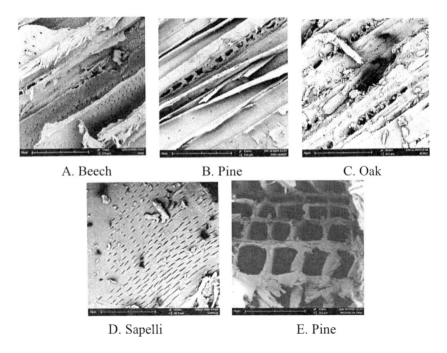

A. Beech   B. Pine   C. Oak

D. Sapelli   E. Pine

Figure 2. Scanning electron micrographs of wood surfaces. A, B, C: 1500x, longitudinal section; D: 5000x, longitudinal section; E: 1500x, cross section.

From this study, it can be concluded that the longitudinal section of wood materials presents a very different structure from the nut shells. In contrast, the cross section of shells and wood presented a similar structure formed with cavities, and consequently, similar mechanical properties should be expected.

Thermal Study: Pyrolysis-GC-MS Analysis

Pyrolysis-GC-MS was applied to study the biomass degradation process of the studied materials. A previous analysis of commercial cellulose and lignin was carried out to identify the main components in these compounds. Figure 3 shows chromatograms obtained for the first pyrolysis cycle, with the main detected compounds indicated in Table 1. No peaks were observed for second or further pyrolysis cycles, indicating that pyrolysis was complete at the conditions used.

Four characteristic compounds were selected for the study of almond shells: 5-hydroxymethyl-2-furancarboxaldehyde and 2-hydroxy-2-cyclopenten-1-one (in cellulose and lignin degradation), as well as vanillin and eugenol (exclusive for lignin degradation). The presence of these compounds in sequential pyrolysis of almond shells was used as a criterion for identification of material degradation.

Table 1. Main compounds detected in commercial cellulose and lignin pyrolysis

Material	Detected compound	Retention time (min)
Cellulose, Lignin	2-hydroxy-2-cyclopenten-1-one	5.5
Cellulose	3-buten-2-one-(dimethylamine)-4-ethoxy	18.2
Cellulose, Lignin	5-hydroxymethyl-2-furancarboxaldehyde	19.4
Cellulose	1,6-anhydro-β-D-glucopyranose	27.1
Lignin	2-methoxyphenol	15.2
Lignin	1-(3-methoxy phenyl) ethanone	21.9
Lignin	Eugenol	23.0
Lignin	Vanillin	24.1

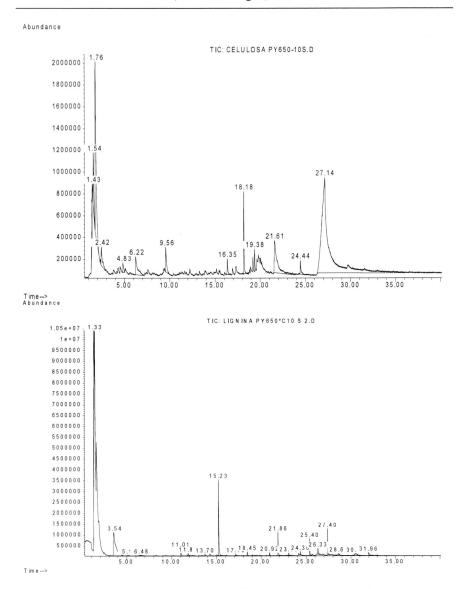

Figure 3. Chromatograms obtained from the first pyrolysis cycle of cellulose and lignin.

The chromatograms obtained from the application of four consecutive pyrolysis processes in almond shells is shown in Figure 4. As can be observed, different peaks appear at low retention times, but their intensity decrease as sequential cycles are applied. The four studied compounds were identified together with other phenolics as 2,6-dimethoxyphenol (22.9 min) and 2,6-dimethoxy-4-(2-propenyl)phenol (31.3 min). The presence of these phenolic compounds has been reported for lignin degradation in wood [12,24]. Peak areas for 5-hydroxymethyl-2-furancarboxaldehyde and 2-hydroxy-2-cyclopenten-1-one (appearing in cellulose and lignin degradation) were higher than those observed for commercial lignin. This fact confirms that both compounds are present in cellulose and lignin degradation. After four sequential pyrolysis processes, only 2-hydroxy-2-cyclopenten-1-one appears, being 5-hydroxymethyl-2-furancarboxaldehyde, vanillin and eugenol present only until the third pyrolysis cycle.

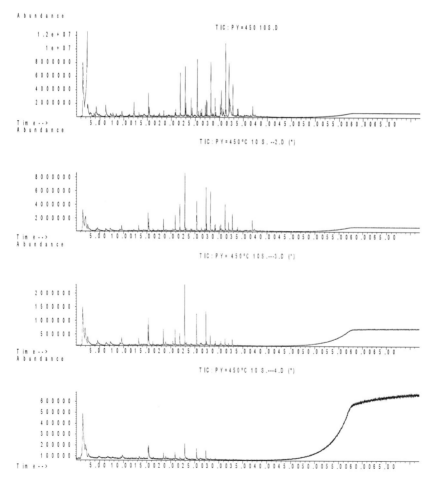

Figure 2. Chromatograms from a pyrolysis sequence (four cycles) of almond shells.

CONCLUSIONS

The SEM morphological study of the longitudinal section of some nut shells and wood has shown their different structure. In contrast, the cross section of shells and woods presented a similar structure formed with cavities and consequently comparable mechanical properties should be expected.

On the other hand, sequential pyrolysis-GC-MS analysis has provided useful information on the degradation process of cellulose and lignin for almond shells. Some volatile compounds, including phenols, furans, alkanes, alkenes, polycyclic aromatic hydrocarbons and ketones were identified in almond shells. In conclusion, the use of lignocellulosic materials from biomass of agricultural wastes could be a reliable possibility for its use in polymer matrices as fillers or blending applications.

ACKNOWLEDGMENTS

Authors would like to thank "Agloma S.L." and "Jesol" for kindly providing the wood and lignocellulosic residuals samples.

REFERENCES

[1] Yang H, Yan R, Chen H, Lee HD, Zheng C, *Fuel,* 2007;86:1781-1788.
[2] Demirbas A, *J. Anal. Appl. Pyrolysis,* 2006;76:285-289.
[3] Williams PT, Nugranad N, *Energy,* 2000;25:493-513.
[4] Lucia LA, *BioResources,* 2008;3:981-982.
[5] Le Digabel F, Boquillon N, Dole P, Monties B, Averous L, *J. Appl. Polym. Sci.,* 2004;93:428-436.
[6] Kadla JF, Kubo S, *Composites A,* 2004;35:395-400.
[7] Stefani PM, Garcia D, López J, Jiménez A, *J. Therm. Anal. Calorim.,* 2005;81:315-320.
[8] Rowell RM, *J. Polym. Environ., 2007*;15:229-235.
[9] Crespo JE, Sanchez L, Parres F, López J, *Polym. Composites,* 2007;28:71-77.
[10] Crespo JE, Sanchez L, Garcia D, López J, *J. Reinforced Plastics Composites,* 2008;27:229-243.

[11] Skodras G, Grammelis P, Basinas P, Kakaras E, Sakellaropoulos G, *Industrial Eng. Chem. Res.* 2006;45:3791-3799.
[12] Hafsi S, Benbouzid M, *Res. J. Appl. Sci.* 2007;2:810-814.
[13] Yanik J, Kornmayer C, Saglam M, Yüksel M, *Fuel Proc. Tech.,* 2007;88:942-947.
[14] Sharma RK, Wooten JB, Baliga VL, Hajaligol MR, *Fuel,* 2001;80:1825-1836.
[15] Theerarattananoon K, Wu X, Staggenborg S, Propheter J, Madl R, Wang D, Transactions of the ASABE *(American Society of Agricultural and Biological Engineers),* 2010;53:509-525.
[16] Bonelli PR, Della Rocca PA, Cerrella EG, Cukierman AL, Bioresource Tech. 2001;76:15-22.
[17] Bonelli PR, *Energy Sources, Part A: Recovery, Utilization, and Environmental Effects*, 2003;25:767-778.
[18] Müller-Hagedorn M, Bockhorn H, Krebs L, Müller U, *J. Anal. Appl. Pyrolysis,* 2003;68:231-249.
[19] Ku CS, Mun SP, *J. Ind. Eng. Chem.* 2006;12:853-861.
[20] Mészáros E, Jarab E, Várhegyi G, *J. Anal. Appl. Pyrolysis,* 2007;79:61-70.
[21] Manya JJ, Velo E, Puigjaner L, *Ind. Eng. Chem. Res.* 2003;42:434-441.
[22] Williams PT, Horne PA, *Fuel,* 1996;75:1051-1059.
[23] Fisher T, Hajaligol M, Waymack B, Kellogg D, *J. Anal. Appl. Pyrolysis,* 2002;62:331-349.
[24] Kuroda K, Nakagawa-Izumi A, *Organic Geochemistry,* 2006;37:665-673.

In: Biodegradable Polymers ...
Editors: A. Jimenez and G. E. Zaikov

ISBN 978-1-61209-520-2
© 2011 Nova Science Publishers, Inc.

Chapter 12

EFFECT OF PROCESSING METHODS ON MECHANICAL PROPERTIES OF SOYA PROTEIN FILMS

P. Guerrero[1], L. Martin[1], S. Cabezudo and K. de la Caba[1]

Universidad del País Vasco. Escuela Universitaria Politécnica
[1]Departamento de Ingeniería Química y del Medio Ambiente
[2]Departamento de Organización de Empresas
Plaza Europa 1. 20018 Donostia-San Sebastián. Spain

ABSTRACT

Glycerol-plasticized soya protein films were prepared through three different methods: solvent casting, compression, and freeze-drying followed by compression. The effects of processing method and glycerol content on the mechanical properties of soya protein films were studied. Young's modulus, tensile strength and elongation at break were evaluated and related to these two variables: processing and plasticizer amount. Results have been explained by using Fourier transform infrared (FTIR) spectroscopy.

Keywords: soya protein, glycerol, processing, films, mechanical properties

INTRODUCTION

Plastics are one of the most important materials employed in daily life. However, environmental pollution from their consumption has become a serious issue, particularly when they are used as one-time use packaging materials. To overcome this problem, biopolymers produced from natural resources are regarded as an attractive alternative since they are abundant, renewable, inexpensive, environmentally friendly and biodegradable [1-3]. Among biopolymers, materials obtained from soya protein are considered potential replacements for petroleum-based products due to their low cost, easy availability and biodegradability [4-6]. Moreover, soya protein is extracted from soybean used to obtain oil. During this process, soya flour is obtained as a secondary product and it can be purified to obtain soya protein concentrates (SPC) and soya protein isolates (SPI), which would add value to agricultural by-products. Although optimization in processing methods is required, this kind of proteins could offer significant opportunities to develop improved packaging materials in the future. These soya-based plastics could be employed as sort-term use or one-time use plastic products in place of the non-biodegradable materials currently used. Once disposed of, soya-based plastic does not take long to biodegrade and it is safer for the environment than more durable petroleum-based plastics. Furthermore, soya protein could be used for food packaging purposes since it meets food grade standards [7].

Soya protein isolate has a protein content higher than 90%, and consists of about 18 kinds of amino acids [8], of which about 62% are polar and reactive amino acid residues [9]. Although soya protein plastics without any additive have a brittle behaviour, which makes processing difficult [10], addition of plasticizers is an effective way to obtain flexible SPI-based films. It is known that plasticizers with molecular characteristics such as small size, high polarity, more than one polar group per molecule, generally impart great plasticizing effect on polymeric systems. Currently, biopolymer films are usually plasticized by hydroxyl compounds [11].

In this paper, we analysed final properties of glycerol-plasticized soya protein films prepared by three different methods: solvent casting, compression, and freeze-drying followed by compression. The effect of glycerol content was also studied. Results were related to data obtained by infrared analysis. The comparison of properties for SPI-based films obtained by different processing methods is an innovative study in this field, in which the most employed technique to prepare films has been dispersion.

EXPERIMENTAL

Materials

Soya protein isolate (SPI, PROFAM 974) with protein content about 90% was supplied by ADM Protein Specialties Division, Netherlands. It was dried for 24 h in an air-circulating oven at 105 °C to determine the moisture content, 5%. Glycerol was food grade reactant obtained from Panreac and it was used without any further purification.

Film Preparation

In order to evaluate relationships between processing and final properties, three processing methods were employed in this study: compression, solvent casting, and freeze-drying followed by compression, as it is illustrated in Figure 1.

In the case of compression, SPI/glycerol mixtures with 70/30, 60/40 and 50/50 (w/w) compositions (designed as SPI30, SPI40 and SPI50) were manually mixed. Mixtures were intensively blended in a beaker for a period of about 5 min, and then they were thermally compacted using a caver laboratory press (Atlas^{TM}). About 1.3 g of SPI/glycerol mixture were placed between 2 sheets of aluminium (0.2 mm thick, 10 mm diameter). These sheets were placed between the plates of the press, which had been previously heated to 150 °C. A pressure of 12 MPa was applied for 2 min. Plates were then allowed to cool for 3 min before removing film samples. Samples were cut to the required gage dimensions for further testing.

In the case of processing by solvent-casting, 7.5 g of SPI were dispersed in 125 mL distilled water and heated at 80 °C and 150 rpm on a magnetic stirrer for 30 min. Furthermore, glycerol was added to the dispersion to obtain SPI30, SPI40 and SPI50 mixtures. These solutions were maintained at 80 °C for other 30 min under stirring at 150 rpm. Subsequently, 40.0 g of each solution were poured into a PS box (14 mm diameter) to obtain approximately the same film thickness (around 70 µm). Samples were kept at room temperature for 48 h to evaporate water. The rest of the sample was freeze-dried using an Alpha 1-4 LD freeze-dryer (Martin Chirst). In this last case, the procedure to obtain films consisted of placing about 1.3 g of mixture between 2 sheets of aluminium as described above.

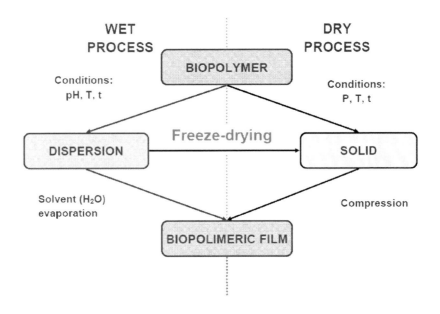

Figure 1. Processing methods: compression, solvent-casting, and freeze-drying followed by compression.

Film Characterization

Fourier transformed infrared (FTIR) spectra of SPI, glycerol and SPI/glycerol films were carried out on a Nicolet Nexus FTIR spectrometer using ATR Golden Gate (Specac). A total of 32 scans were performed at 4 cm^{-1} resolution. The measurements were recorded between 4000-400 cm^{-1}.

Tensile strength and elongation at break were determined according to ASTM D1708-93 with a Minimat 2000 (Rheometrics Scientific). Films were conditioned in a controlled environment chamber (Dycometal CCK81) at 25 °C and 50% RH for 48 h before testing. Film thickness was measured to the nearest 0.001 mm with a hand-held micrometer (Mitutoyo). Three thickness measurements were taken on each specimen at five different positions, being in the range of 60-80 μm in all cases.

RESULTS AND DISCUSSION

FTIR Analysis

SPI employed in this study contents mainly acidic amino acids, glutamic acid (19.2 wt%) and aspartic acid (11.5 wt%), and approximately 1 wt% of cystine, so thus its isoelectric pH is acid, 4.6 for PROFAM 974.

The FTIR spectrum of pure SPI is shown in Figure 2. The main absorption peaks are related to C=O stretching at 1630 cm^{-1} (amide I), N-H bending at 1530 cm^{-1} (amide II) and C-N stretching (amide III) at 1230 cm^{-1} [12]. The broad band observed in the 3500-3000 cm^{-1} range is attributable to free and bound O-H and N-H groups, which are able to form hydrogen bonding with the carbonyl group of the peptide linkage in the protein [13].

The FTIR spectral data of pure glycerol are show in Figure 3. Typical absorption bands of glycerol are located in the region from 800 cm^{-1} up to 1150 cm^{-1}, where five peaks corresponding to the vibrations of C-C and C-O linkages appear. The peaks at 850 cm^{-1}, 925 cm^{-1}, and 995 cm^{-1} are assigned to the vibration of the C-C skeleton, the band at 1045 cm^{-1} is associated to the stretching of the C-O linkage in C1 and C3, and the one at 1117 cm^{-1} is related to the stretching of C-O in C2.

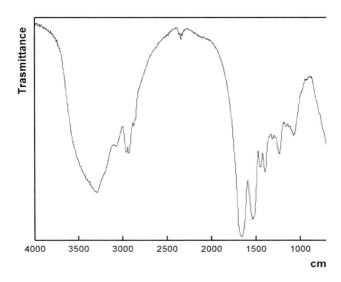

Figure 2. FTIR spectrum of pure SPI used in this study.

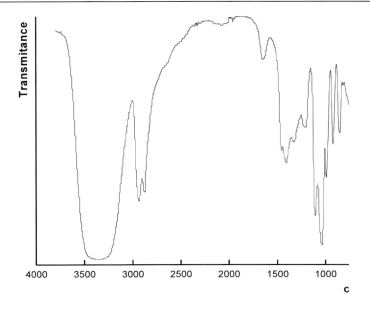

Figure 3. FTIR spectrum of glycerol.

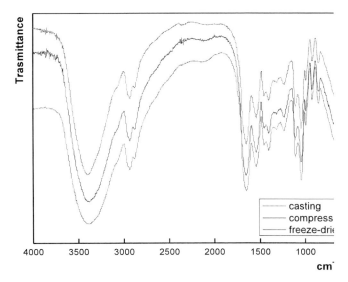

Figure 4. FTIR spectra of SPI40 films processed by solvent-casting, compression and freeze-dried followed by compression.

Effect of Processing Methods on Mechanical Properties ... 185

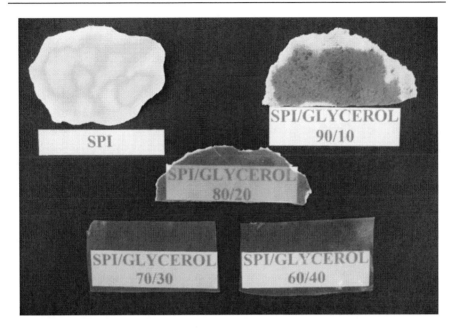

Figure 5. Visual aspect of SPI-based films obtained by compression with different glycerol contents.

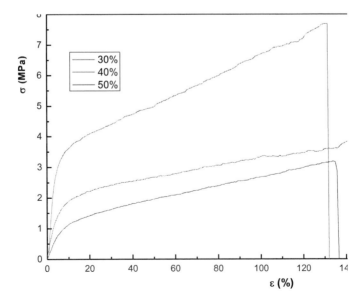

Figure 6. Effect of glycerol content on tensile strength in films based on SPI processed by compression.

Figure 4 shows spectral data for SPI plasticized with 40 wt% glycerol. Comparing this spectrum with the previous ones corresponding to pure SPI and glycerol, it can be observed that no changes take place in the characteristic peaks of both protein and glycerol. In particular, there is no change in the 1200-800 cm^{-1} region; the bands at 850 cm^{-1}, 925 cm^{-1}, 995 cm^{-1}, 1045 cm^{-1} and 1117 cm^{-1} appear for SPI40 system in the same position as in Figure 3. This fact indicates that glycerol does not react with the protein through covalent linkages. In the case of those films obtained by solvent-casting, a broader peak corresponding to the hydroxyl group could be observed, indicating a major moisture content in the films processed in solution.

MECHANICAL PROPERTIES

The films with glycerol content lower than 30 wt% were fragile and it was no possible to cut samples for mechanical analysis, as it is shown in Figure 5. On the other hand, the films obtained with 50 wt% glycerol were very sticky and also unable to be cut for mechanical testing. However, films obtained with 30 and 40 wt% glycerol were flexible and they presented good mechanical properties. In these cases, glycerol acts as a plasticizer without forming covalent linkages with SPI, as it was shown by FTIR analysis. However, due to the three hydroxyl groups present in glycerol, it is expected that it could be be strongly bonded by hydrogen bridges with protein molecules at amine, amide, carboxyl and hydroxyl sites. Being small in size, glycerol effectively increases the free volume of the system.

The hydroxyl groups of glycerol could interact with amino and acid groups in the protein, decreasing inter- and intramolecular interactions between protein chains, such as hydrogen bonds, and improving the motion ability of protein macromolecules, resulting in the flexibility of materials. As the result, the behaviour of films based on SPI changes from brittle to flexible, as it can be observed in Figure 6.

Soya proteins have polar and non-polar side chains, which promote strong inter- and intramolecular interactions, such as hydrogen bonding, dipole-dipole, and so on. The strong charge and polar interactions between side chains of soya protein molecules restrict segment rotation and molecular mobility, leading to an increase of modulus and tensile strength. According to Krochta [14,15], two different events occur during the film formation. Firstly, during the heating phase, protein structure is disrupted, some native disulfide bonds are cleaved, and sulfhydryl and hydrophobic groups are exposed. Then, during the film drying phase, new hydrophobic interactions occur and also new hydrogen bonds are formed.

There are several typical inter- and intramolecular interactions, such as hydrogen-bond, disulfide-bond, dipole-actions, charge-charge, and hydrophobic interactions, in soya protein that are characteristic of natural proteins. According to the amino acid composition of SPI, hydrogen bonding occurs among $-NH_2-$ (in arginine and lysine), -NH- (in proline and histidine), -OH (in tyrosine, threonine, and serine), -COOH (in glutamic acid), and peptide bonds. It seems that the density and strength of the interactions are greatly different at specific locations in SPI molecules and, as a result, soya protein molecules contain different regions with distinct abilities to accept glycerol molecules [16].

The increase in the amount of glycerol causes a decrease of tensile strength and an increase of elongation at break due to the fact that glycerol reduces interactions between protein chains, and consequently increasing the chain mobility. Nevertheless, when the amount of glycerol added is 50 wt% no further improvement of elongation is observed. Taking the above in consideration, the effect of the processing method in the mechanical properties of films was studied for the SPI40 system, and results are shown in Figure 7.

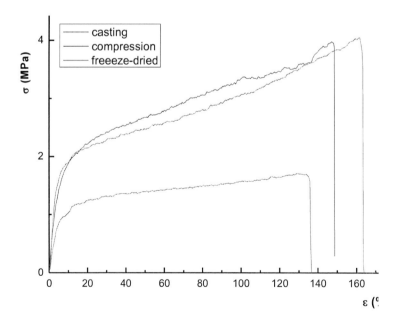

Figure 7. Effect of the processing method for SPI40 system.

Films obtained by compression, as well as those freeze-dried and further pressed, showed similar tensile strength and elongation at break. No differences were found between the films obtained by manual mixing and those obtained by dispersion in water while heating, so it can be said that the diffusion of glycerol in the protein is independent from the system employed to mix the protein and the plasticizer. In contrast, films obtained by solvent-casting showed the worst mechanical behaviour, in other words, the lower resistance and elongation at break.

CONCLUSIONS

The three hydroxyl groups in glycerol can form hydrogen bonds with amine, amide, carboxyl and hydroxyl groups of the protein, increasing the free volume in the system and causing changes in the behaviour of the material from fragile to flexible. The flexibility of films obtained increased when the glycerol content increased up to 40 wt%, indicating the efficient plasticizing effect of glycerol. However, higher glycerol contents did not improve elongation at break, which indicates that glycerol would be in excess. The use of plasticizers to break intermolecular linkage, which stabilizes the protein in their primitive structure, makes the protein chains mobile. The orientation and restructuring of chains as well as the formation of new intermolecular links stabilize the three-dimensional network formed.

ACKNOWLEDGMENTS

The authors thank Diputacion Foral de Gipuzkoa (35/08), the University of the Basque Country (EHU08/26), and the Basque Government (S-PE08UN08) for their financial support.

REFERENCES

[1] Bastioli C. *"Handbook of biodegradable polymers"*, Shropshire; Smithers Rapra Technology, 2005.
[2] Belgacem, MN, Gandini, A. *"Monomers, Polymers and Composites from Renewable Resources"*, Oxford; Elsevier, 2008.

[3] Yu L. *"Biodegradable Polymer Blends and Composites from Renewable Resources"*, Oxford; Elsevier, 2009.
[4] Gennadios A, Hanna, MA, Kurth LB. *Food Science and Technology-Lebensmittel-Wissenschaft and Technologie,* 1997;30:337-350.
[5] Gennadios A. *"Protein-based films and coatings"*, Florida; CRC Press, 2002.
[6] Kumar R, Choudhary V, Mishra S, Varma, IK, Mattiason, B., *Industrial Crops and Products,* 2002;16:155-172.
[7] Damodaran S, Paraf A. *"Food proteins and their applications"*, New York; Marcel Dekker, 1997.
[8] Zhong, Z, Sun, SX. *Polymer,* 2001;42:6961-6969.
[9] Lodha P, Neteravali A N. *J. Mat. Sci.*, 2002;37:3657-3665.
[10] Feng L, Wang, YQ, Sun, XS. *J. Polym. Eng.* 1999;19:383-393.
[11] Cao N, Yang X, Fu Y. *Food Hydrocolloids,* 2009;23:729-735.
[12] Schmidt V, Giacomelli C, Soldi V. *Polym. Degrad. Stabil.,* 2005;87:25-31.
[13] Karnnet S, Potiyaraj P, Pimpan V. *Polym. Degrad. Stabil.,* 2005;90:106-110.
[14] Sothornvit R, Krochta JM. *J. Food Eng.,* 2001;50:149-155.
[15] Sothornvit R, Olson CW, McHugh TH, Krochta JM. *J. Food Eng.* 2007;78:855-860.
[16] Knubovents J, Osterhout JJ, Connolly PJ, Klibanov AM. *Proceedings of the National Academy of Sciences USA,* 1999;96:1262-1267.

In: Biodegradable Polymers ...
Editors: A. Jimenez and G. E. Zaikov

ISBN 978-1-61209-520-2
© 2011 Nova Science Publishers, Inc.

Chapter 13

DEVELOPMENT OF A BIODEGRADABILITY EVALUATION METHOD FOR LEATHER USED IN THE FOOTWEAR INDUSTRY

M. A. de la Casa-Lillo^{2,*}, A. Diaz-Tahoces^{2}, P. N. de Aza-Moya^{2}, P. Mazón-Canales^{2}, V. Segarra-Orero^{1}, M. A. Martínez-Sanchez^{1} and M. Bertazzo^{1,†}

[1]INESCOP – Footwear Technological Institute, Polígono Industrial "Campo Alto", P.O. Box 253, 03600 Elda (Alicante), Spain
[2]Instituto de Bioingeniería. Miguel Hernández university of Elche, Avenida de la Universidad s/n, 03202 Elche (Alicante), Spain

ABSTRACT

This article aims to contribute to find of a solution to one of the main environmental problems in the footwear industry: the large quantity of waste which is produced and the long life-cycle of the materials due to the use of chromium compounds in tanning processes. Alternative tanning methods are being studied which allow a faster degradation rate of the materials used. In ordert to face this situation, it is first necessary to

---

* E-mail: mcasa@umh.es
† E-mail: mbertazzo@inescop.es

develop a quick method for the determination of the biodegradability to be able to foresee which of these products would be more biodegradable. INESCOP, together with the Miguel Hernández University of Elche (UMH), is developing a method based on the principle of natural biodegradation of organic compounds by microorganisms' action. The final objective is to get a method for a quick measurement of biodegradability of leather tanned with different industrial tanning methods, and to be able to establish which process should be used in order to achieve a lower environmental impact.

Keywords: Biodegradability, aerobic, leather, standardisation, footwear industry

INTRODUCTION

The tanning process transforms leather into a resistant, long-wearing and useful material with a wide variety of applications, among them those related to footwear and leather goods stand out due to their market sales volume. In a general view of the leather industry it can be said that production is split between ten countries on a world scale, with production figures adding up to more than 18,000 million feet^2 in 2000 [1]. With respect to tanned leather production, Europe is at the head with 25% leather production, generating almost 170,000 tons of leather waste per year, 43,000 tons of which are produced by the footwear sector. A factory producing around 100 tons of leather daily produces 30 tons of this waste every day, which turns tanned leather waste management into a serious environmental problem. The implementation of clean technologies in the leather sector would greatly reduce the pollutant load generated by tanneries [2,3].

Furthermore, although the pollution generated by this industry is an important issue it is not the only one, since the management of the solid waste generated by the discards from footwear, leather goods and textiles which all end up in dumping sites at the end of their useful lives is also significant. This makes the waste management system very inactive due to the non-biodegradability of such waste. The main environmental problem that this issue brings about is the great quantity of long-life waste produced due to, amongst other factors, the use of chromium compounds in the leather tanning process. During this process the leather is transformed into a product that keeps its fibrous structure interlinking the collagen fibres by means of chemical bonds. Once these bonds have been formed the leather is stable to bacterial attack making the final product, leather, a material which is very resistant to biodegradation. Considering the fact that the average life of tanned leather is 25-40 years, the accumulation of this waste in dumping sites and the subsequent management of its disposal proves imply a high economical and environmental cost. This fact highlights the great importance of finding a method to minimise the problem. The use of materials in leather production which would allow the post-treatment of solid waste, both industrial and urban, with biodegradation techniques, would mean a significant reduction or indeed the complete elimination of this type of waste.

Within this context it is important to underline the fact that there currently does not exist a specific standard for determining the biodegradability of leather in aerobic conditions. With this in mind, this paper looks to find a solution for the need to establish a method with which the various production methods of the industry could be evaluated, so that once understood, the environmental impact due to the aforementioned accumulation of waste could be reduced.

There is a general consensus that a material can be defined as biodegradable in the case the degradation is due to the action of microorganisms and the final result is that the material is turned into water, carbon dioxide and/or methane and new biomass [4]. Therefore, the main objective of this work is to develop a method which would make possible the measurement of biodegradability of leather and following this, would give an estimation of its level regarding the different tanning systems which are used today, in this way providing a solution to the important sector demand in this respect.

MATERIALS AND METHODS

Previous to carry out experiments, biodegradability standards of different materials existing in the literature were studied [5-11]. No biodegradability standards concerning leather were found and for this reason standard ASTM D 5209-92 [5] which refers to plastics was chosen as a reference. One factor which contributed to this selection is the fact that both plastic and leather share some properties and components. The ASTM D 5209-92 standard describes the aerobic biodegradation methods of plastic materials in the presence of municipal sewage sludge (MSS). By measuring the indirect levels of CO_2

produced and taking into account the organic carbon content available, the materials' level of biodegradability was determined.

In these tests a synthetic medium made up of 6 mL phosphate buffer, 8.5 g KH_2PO_4, 21.45 g K_2HPO_4, 33.4 g $Na_2HPO_4 \cdot 7H_2O$, 1.7 g NH_4Cl, 12 mL of iron chloride (0.25 g L^{-1}), 3 mL of ammonium sulphate (40 g L^{-1}), 3 mL of magnesium sulphate (22.5 g L^{-1}), 3mL of calcium chloride (27.5 g L^{-1}), 2470 mL of distilled water (with a carbon content of < 2 mg L^{-1}) was used. Tests were performed at a temperature 23 °C ± 1 °C and throughout the whole process the reaction flasks were constantly shaken.

The elemental analysis of carbon present in each sample was carried out using a Thermo Finnigan Flash EA series 1112 equipment. The calculation of total carbon initially present in each sample was used to calculate the maximum amount of CO_2 which could have been generated during the biodegradation tests.

The tests were carried out in the presence of two controls, a positive one made up of a synthetic medium, microorganisms and a standard, and a negative one, made up only of the synthetic medium and an inoculum. As the inoculum could theoretically have organic waste apart from microorganisms, the CO_2 values obtained during the tests in these flasks were a consequence of the CO_2 values obtained in the testing flasks and of those of the positive control.

For the tests on leather some modifications were made to the D 5209-92 standard [5] due to the need of adapting the method to fit with the requirements of the samples. The main modifications introduced to the system were the use of collagen Type I (Sigma) as a standard and the percentage (5%) and type of inoculum (tannery wastewater or TWW) used.

For the biodegradability studies samples were tanned cattle hides treated with different tanning agents such as phosphonium (THPs), oxazolidine (Oxasoft), chromium and resins. All samples were introduced in powder form. The quantity of carbon dioxide produced by the aerobic digestion of samples by microorganisms was calculated by measuring the barium hydroxide carbonation (0.025N) in the analysis flasks.

A direct determination of the biodegradation can be performed by analysing the weight loss of samples. This value was determined by weighing the sample which was introduced in each testing flask and comparing with the final weight obtained after filtering and drying the material once the experiment had been concluded.

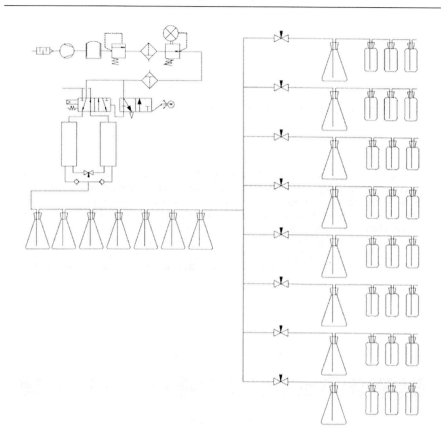

Figure 1. Schematic diagram of the biodegradability unit.

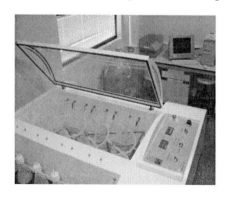
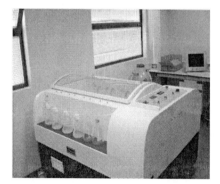

Figure 2. Biodegradability equipment.

$$\%weightloss = \frac{start\ weight - final\ weight}{start\ weight} \times 100$$

For leather biodegradation tests a prototype was designed (Figure 1) allowing the analysis of 8 samples per test. Culture methods, testing temperature and CO_2 quantification were the same as those described in ASTM D 5209-92 standard [5].

This is a compact unit whose design (Figure 2) is based on standard ASTM 5209-92 [5]. In general, its application can be extrapolated to study the biological degradation of any textile and/or polymeric material. Therefore, the experimental method covers the determination of the degree and rate of the material's degradation under controlled conditions, which allows the analysis of the CO_2 produced during the tests. The unit complies with strict requirements in terms of flow, temperature and agitation control.

The equipment features an autonomous clean air (CO_2 free) generation system comprising a silent compressor connected to a PSA (pressure swing absorption) with a molecular sieve filter provided by the Peak Scientific PSA model PG14L with a flow 17 L min^{-1} which is the time necessary to ensure that a stable concentration of CO_2 not lower than 1 ppm is achieved in the air and leaving the generator through the PSA for 16 hours.

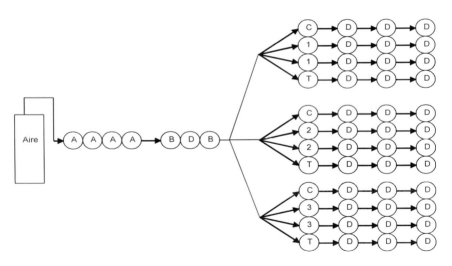

Figure 3. Diagram of the method proposed in the ASTM D5209-92 standard.

The prototype works in such a way that the air generated bubbles through an array of seven flasks (pre-treatment flasks) which trap the residual carbon dioxide of the air flow coming from the PSA to be later divided in eight lines controlled by eight valves which permit individual control of each flow and which in turn feed eight flasks (reaction flasks) located inside the tank. The outlet of each flask is directly connected to an array of 3 more flasks (analysis flasks) from which the results will be obtained (figure 3).

RESULTS AND DISCUSSION

Results were obtained by using minimal culture medium and tannery wastewater (TWW) as inoculum.. All the values shown in the kinetic biodegradation curves were corrected by the negative control which was composed only of the inoculum and the culture medium.

The skin is made up of 3 main parts: the epidermis or corneous extract; the dermis (corium or cutis) with ca. 85% of total thickness of raw skin and finally the hydrodermis or subcutis. The chemical composition of leather is mainly carbon (45-55%); hydrogen (6-8%); oxygen (19-25%); nitrogen (16-19%) and other elements like sulphur, phosphorus, iron, bromine and chlorine (0.5-2.5%) [12].

The chemical composition of the samples used for the different tests can be seen in Table 1. Elemental analysis allows calculating, from the percentage of carbon present in each leather sample, the maximum amount of CO_2 that could be produced during the biodegradation process of leather. Based on these values, the biodegradation percentage can be determined.

Table 1. Elemental analysis of the leather samples

Samples	Nitrogen %	Carbon %	Hydrogen %	Sulphur %	Theoretical CO_2 (g)
Collagen Sigma (Type I) (positive control)	12.82	48.14	7.25	0.00	0.8855
Collagen + Phosphonium	10.44	47.35	6.43	0.77	0.8681
Collagen + Oxazolidine	11.40	42.85	6.84	1.32	0.7856
Collagen + Chromium	9.67	44.60	6.86	1.22	0.8177
Collagen + Vegetable Resin	11.53	45.59	6.73	1.91	0.8358

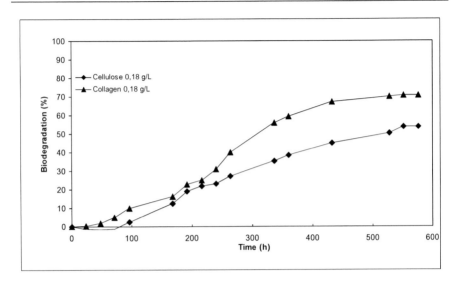

Figure 5. Comparison between the biodegradability of collagen and cellulose.

The positive standard used in all of the biodegradation tests was defined by directly comparing the collagen (Type I, Sigma) and cellulose (Sigma).

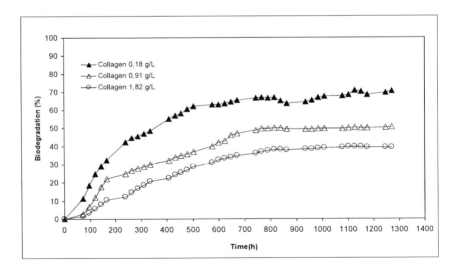

Figure 6. Biodegradation with different initial collagen concentrations.

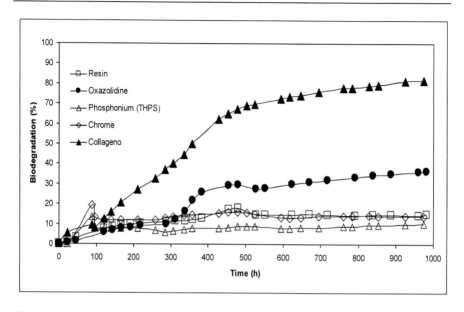

Figure 7. Biodegradation of hides (% CO_2 produced).

As can be observed in Figure 5, after 580 hours of testing, the collagen biodegradation increased about 70% and it was approximately 15% higher than the degradation observed in cellulose. Based on these results, collagen was chosen as a standard substrate since it proved to be more degradable and it is the natural constituent of the leather used in the footwear industry.

Once the collagen was defined as the standard substrate, some tests were carried out to determine the optimum concentration which should be used. Three different initial collagen concentrations were used: 0.18 g L^{-1} (0.5 g); 0.91 g L^{-1} (2.5 g) and 1.82 g L^{-1} (5.0 g). As can be observed in figure 6, all curves showed similar collagen degradation profile with a short lag phase followed by an exponential phase of approximately 500 hours until the stationary phase between 600 and 700 hours was reached. The main difference between curves was in the final collagen biodegradation percentage. Thus, an initial collagen concentration of 0.18 g L^{-1} led to 70% biodegradation after 45 days. In the same period, the degradation level of the other concentrations tested (0.91 and 1.82 g L^{-1}) were, respectively, 50 and 35% of the initial level. Results showed that the lower the initial concentration of collagen used, the higher the final percentage of biodegradation obtained. This difference could be related to the fact that for higher substrate concentrations a longer testing time was needed and therefore higher biodegradation levels were attained. To verify if this hypothesis was correct the test was extended to 1320 hours (55 days), and no increase in the degradation percentage was observed between the different collagen concentrations tested (Figure 6). After 1320 hours, 18 g L^{-1} collagen concentration still showed the optimum level of degradation.

Based on these results it was observed that the lower the initial concentration of the substrate, the higher the final percentage of biodegradeation achieved, with a concentration of 0.18 g L^{-1} established as the optimum value. Concentrations lower than the optimum were discarded since the quantification method used presented difficulties in measuring (data not published).

Once the optimum initial substrate concentration was determined a series of tests were performed on hides treated with different tanning agents (Figure 7).

Samples were then treated with vegetable resin, oxazolidine, phosphonium and chromium. A sample of pure collagen (Sigma, Type I) was used as a positive control. The principle of the method for the evaluation of leather biodegradation is based on the indirect measurement of CO_2 levels produced by the microorganisms during the degradation process using $Ba(OH)_2$. Each analysed hide showed a quantity of carbon which was determined by the elemental analysis of these hides and served for the calculation of the theoretical CO_2 that can be generated by each hide.

As expected, pure collagen showed the highest biodegradation rates, approximately 80% in 1000 hours. In the same period, hides treated with chromium, the most common tanning agent, presented a biodegradation percentage of 12%. This value was quite similar to those observed in phosphonium-tanned leather which was 9%; and vegetable resin-tanned leather (14%). These results prove that most of the leather that had been tanned with alternative tanning agents did not provide results significantly different when compared to chromium, and therefore they did not provide an adequate alternative for its replacement in the tanning process. On the other hand, samples treated with oxazolidine showed biodegradation percentages up to 35%. This value is 21% higher than the best result referred earlier for hides treated with vegetable resin and 26% higher than that observed in phosphonium-treated hides. This result, although still far from the 83% of degradation observed with pure collagen, represents a significant improvement in comparison to chromium and it shows the possibility of an alternative tanning agent for leather used in the footwear industry.

Table 2. Biodegradation of leather (% weight lost)

Sample	(% weight loss)
Collagen	83.1 – 89
Phosphonium	10.8
Chromium	12.5
Oxazolidine	34
Resin	14

The results obtained in the biodegradation tests were compared to the percentage weight loss measured in the samples used in the tests. After each test all contents that were not degraded were filtered, dried and weighed and weight loss was estimated (Table 2).

Those values were very similar to those obtained in biodegradation tests which substantiates the validity of this new testing method with the advantage that the analysis by dry weight can only be achieved at the end of the test and that all the information obtained in the biodegradation process, in the case of the dry weight, would have been lost.

CONCLUSIONS

The simulation in vitro of a natural process is a great challenge for researchers due to the real complexities of natural phenomena. This fact enforces the necessity of using the largest number of variables possible in the work carried out in laboratories, which almost always results in the reduction of the possibilities to get reproducible and easily interpretable tests. In this way, a good evaluation method is that which refers to simple experimental methodologies that are close to the real phenomena which occur in Nature. The method developed in this work does not reflect all the variables which can be observed in Nature in leather biodegradation but achieves in a simple and easily interpretable way an estimation of the biodegradation level of the hides submitted to alternative tanning methods by indirectly quantifying the CO_2 produced by their biodegradation by aerobic microorganisms. In this way it can be determined that leather in its pure form (collagen) shows biodegradation values of up to 90% in a period of no longer than 40 days. This value is significantly higher than those observed in all the tested samples. Phosphonium and vegetable resin-tanned leather resulted in degradation not higher than 15% in the same testing time, values similar to those observed in chromium-tanned hides, and therefore they cannot be considered as alternative tanning methods. Samples treated with oxazolidine, however, showed biodegradation values of up to 34% which is 2.5 times higher than the chromium treated hides. Although these values are far lower than those observed in collagen, oxazolidine could be considered to be a possible substitute for chromium in the tanning process.

Acknowledgments

The authors thank IMPIVA and the European Regional Development Fund (ERDF) for their financial support.

References

[1] *Textile and Production Centre* (CITYC). Bulletin No. 7, September 2005.
[2] Dhayalan K, Nishad Fátima N, Gnanamani A, Raghava Rao J, Unni Fair B, Ramasami, T. *Waste Management,* 2007;27:760-767.
[3] Simeonova LS, Dalev PG. *Waste Management,* 1996;16:765-769.
[4] EN 13432:2000: "Requirements for packaging recoverable through composting and biodegradation- Test scheme and evaluation criteria for the final acceptance of packaging".
[5] ASTM D5209-92: "Standard Method for determining the aerobic degradation for the plastic materials in the presence of municipal sewage sludge".
[6] ASTM D5210-92: "Standard Method for determining the anaerobic degradation for the plastic materials in the presence of municipal sewage sludge".
[7] ASTM D5338-98 (2003): "Standard Method for determining the aerobic degradation of plastic materials under controlled composition conditions".
[8] ISO 14851:1999: "Determination of the ultimate aerobic biodegradability of plastic materials in an aqueous medium - Method by measuring the oxygen demand in a closed respirometer".

[9] EN ISO 14852:2004: "Determination of the ultimate aerobic biodegradability of plastic materials in an aqueous medium - Method by analysis of evolved carbon dioxide".
[10] ISO/DIS 14853:2005: "Determination of the ultimate anaerobic biodegradability of plastics materials in an aqueous medium. Method by measurement of biogas production".
[11] EN ISO 14855:2004: "Determination of the ultimate aerobic biodegradability and disintegration of plastic materials under controlled composting conditions - Method by analysis of evolved carbon dioxide".
[12] http://www.basf.com/leather. Pocket Book for the Leather Technologist. Last modified: Jan 22, 2007.

In: Biodegradable Polymers ...   ISBN 978-1-61209-520-2
Editors: A. Jimenez and G. E. Zaikov   © 2011 Nova Science Publishers, Inc.

Chapter 14

CHROMIUM TANNED LEATHER WASTE ACID EXTRACTION, RESIDUE RECYCLING AND ANAEROBIC BIODEGRADATION TESTS ON EXTRACTS

Maria J. Ferreira[1,*], Manuel F. Almeida[2], Vera Pinto[1], Isabel Santos[1], José L. Rodrigues[1], Fernanda Freitas[1] and Sílvia Pinho[2]

[1]Centro Tecnológico do Calçado Portugal, R. Fundões, Devesa Velha, 3700-121 São João da Madeira, Portugal
[2]Laboratory of Processes, Environment and Energy Engineering, Engineering Faculty of Porto University, R. Dr. Roberto Frias, 4200-465 Porto, Portugal

ABSTRACT

Chromium sulfate tanned leather wastes are currently mainly landfilled, despite their content in valuable biopolymers and minerals. A more sustainable option might be recovering as much as possible of its chromium and, consequently, lowering its content in the resulting leather scrap, thus facilitating the recycling of the remaining material. With this

---

[*] Corresponding author: Maria J. Ferreira, email: mjose.ferreira@ctcp.pt

objective, chromium leather scrap was leached with sulfuric acid solutions at low temperature (296 and 313 K) aiming at maximizing chromium removal with minimum attack to the leather matrix. Although reasonable Cr recoveries were achieved (30 and 60 %), chromium in eluate from leaching the residue according to DIN 38414-S4:1984 standard exceeded the threshold value, being considered hazardous. Thus, it was methodically washed with water and alkaline solutions in order to remove or stabilize the chromium de-linked from collagen. Furthermore, the so-treated leather scrap was recycled in rubber compounds for shoe soles application, confirming its potential to improve tear resistance. However, fine tune of the formulation is necessary to avoid failure due to poor tensile strength. The anaerobic biodegradation of the acid solution resulting from Cr recovery was evaluated indicating anaerobic biodegradability in the range of 25 to 45 %.

Keywords: Leather waste, de-chroming, anaerobic biodegradation, rubber composites

INTRODUCTION

Chromium-tanned leather shavings and finished leather scrap are normally disposed of in landfill sites. As these wastes have low density and present low compaction ability, they have high landfilling costs per mass unit, which may be substantially increased when they are considered hazardous.

Since the main reason for chromium-tanned leather wastes environmental concern is the presence of chromium, for the last 30 years, research has been directed towards finding tanning alternatives with less environmental impact, unfortunately without significant results due to the unique properties that chromium confers to the resulting products. In the case of shoe production, still 70-80 % of the leather used is chromium-tanned due to superior physical, mechanical and aesthetic properties as well as easy industrial processing.

Several options have been studied to give added value to the materials contained in leather scrap, mainly protein (collagen) and chromium, namely by: i) producing smaller items; ii) making leather board, materials "similar" to leather and thermal and acoustic insulation plaques [1,2]; iii) preparing leather powder for the adsorption of chemicals (dyes and chromium) [3-6]; iv) using in combustion, incineration, gasification or pyrolysis to produce energy and make by-product chemicals, including activated carbon and chromium salts [7-11]; v) hydrolyzing them to recover proteins and chromium.

From an ecological point of view the use in smaller parts and recycling in new materials are the preferred options. However, these systems have their limitations and they are not able to deal with the huge amount of leather materials generated annually. In the case of shoe industry, it can be estimated that an average of 0.1-0.2 kg of finished leather waste is generated by pair produced. In the EU footwear industry, near 1000 million pairs were produced during 2007 [12], thus generating about 150,000 metric tonnes of this waste. Furthermore, the presence of chromium may limit the materials application, namely to prevent the spread of heavy metals.

Destruction processes, such as hydrolysis and incineration have been proposed for chromium leather treatment. The first one is a wet process using acids, oxidants, bases, enzymes, pressure or complexing agents, and presents some advantages over thermal processes, namely by minimizing production of polychlorinated phenols, dioxins and other organic halogens at higher temperatures [13-15].

Acid treatments, with sulfuric acid or other complexing agents, namely formic acid, oxalic acid, organic oxiacids, or sodium sulfate, alone or in combination, in conditions ranging from mild de-chroming up to complete attack of the leather structure, are able to remove and dissolve chromium(III) from the chromium-collagen complex [16-19]. Sulfuric acid hydrolysis has been used to convert wet blue shavings into an hydrolyzate containing highly masked chromium besides free anions from collagen demolition. Skins treated with this product present greater fullness, improved feel and brighter color [20,21]. Shavings hydrolysis at higher temperatures (366 K) result also in hydrolyzates usable as retanning agent namely for split leather enhancing its physical properties [22]. Treatment of shavings with formic acid (3 h at 368 K) followed by modification of hydrolyzates by vinyl monomers to produce a retanning and filling agent has also been reported [23]. More recently, formic, phosphoric and nitric acids were used to treat products resulting from chromium shavings alkaline enzymatic hydrolysis [24].

The application of milder hydrolysis conditions permits chromium separation from the detanned waste by filtration and the collagen protein obtained remains more or less intact and has low chromium content, thus facilitating its recycling [19]. The dissolved chromium might be usable as such or preferably after purification as complement to the tanning or retanning agent.

The present work applies a wet acid treatment to finished chromium tanned leather scrap aiming at recovering chromium with limited organic matrix attack, in a near static low cost treatment. That industrial waste is originated after the tanning, retanning, fatliquoring and finishing steps at the leather production sector as well as during leather products manufacturing. Hazardousness of leather scrap and final residue was assessed as well as the anaerobic biodegradability of the resulting extraction solutions. These solutions containing chromium may be used as retanning agent by following classical methodologies established in the literature [25,26] and tested by the authors [27,28], thus closing the metal cycle. One of the uses for the partially detanned leather resulting from wet acid treatment is the introduction in vulcanized rubber. Leather in rubber formulations is prone to affect the vulcanization characteristics and vulcanizates properties due to reactive functional groups present, trivalent chromium content and acidic nature. Procedures for neutralizing the acidic nature of the leather may be employed before they are added to rubber. Ravichandran et al. [29,30] studied scrap rubber recycling in natural rubber (NR) using untreated and neutralized fibrous particles of leather shavings as processing aid. To overcome the acidic nature, 1% solution of urea, aqueous ammonia and sodium bicarbonate were used separately as neutralizing agents. Neutralized leather shavings were also studied as filler for nitrile rubber (NBR) reinforcement. Shavings neutralized with Na_2CO_3 and NH_3 showed improved vulcanization characteristics and mechanical properties, whereas shavings neutralized with NaOH exhibited poor properties [31]. Shavings of chromium-tanned wastes have been used to fill isoprene rubber (IR), carboxylated acrylonitrile butadiene rubber (XNBR) and NBR, and the obtained vulcanizates were reported as having good usable properties [32]. Chromium-tanned fatliquored leather buffing dust blended with five parts by weight of ZnO (renders dust surface more alkaline and favoring dispersion and vulcanizates properties) was used as a filler for XNBR and NBR, improving their physical and mechanical properties [33].

These recent studies on leather fibers showed that production of environmentally-friendly composite materials with leather wastes is feasible. Therefore, another goal of the present work was to screen the use of completely finished chromium-tanned leather waste fibers partially detanned and neutralized (using H_2O, NH_3 and Na_2CO_3) as reinforcing additives and fillers for shoe components. For this purpose, a comparative study on the effect of incorporating the treated fibers (grinded to ≤1mm) in different quantities (15 to 25 parts per hundred parts of rubber (phr)) on physical, mechanical and chemical properties of NBR has been carried out.

EXPERIMENTAL

Leather Sample Preparation and Characterization

The de-chroming trials used six finished chromium-tanned leathers presenting aesthetics, touch and general properties usual in shoe manufacture. They were cut to 5 x 3 cm^2 pieces, then homogenized and conditioned in standard laboratory atmosphere (296 ± 2 K and 50 ± 5% relative humidity). The 5 x 3 cm^2 size was chosen to simulate an average scrap size, easily obtainable at the industrial plants. For chemical characterization and leaching tests, this material was shredded to ≤4 mm (using a Pegasil®-Zipor® mill with rotating knives and a 4 mm sieve), thoroughly homogenized and conditioned as above indicated.

The leather sample was chemically characterized by following the standards listed in Table 1 [34-54] and results are presented in Table 2. The material was subjected to leaching tests according to EN 12457-2:2002 standard [42], with liquid to solid ratio of 10 L kg^{-1}. This test is similar to the DIN 38414-S4:1984 [53] recommended by Portuguese law D.L. 152/2002 [51]. Samples and the respective eluates were characterized regarding chromium content by following standards listed in Table 1 and the results are also presented in Table 2.

Leather De-chroming Tests

The chromium recovery tests were carried out in triplicate by following conditions previously found as appropriate [16]: (i) leather/solution ratios of 1/5 or 1/10; (ii) solutions with 25 mL concentrated H_2SO_4 L^{-1}; (iii) 72 + 72 h for time of contact; and, (iv) holding temperature of 296 or 313 ± 2 K with occasional shaking (manual shaking with 20 revolutions each 12 hours), respectively. The obtained materials/slurries were vacuum filtered through Buchner glass funnels using Whatman n.1 filter paper and thoroughly washed with distilled water to remove the extraction solution. This washing solution was added to the previously separated extraction solution. The residue was further washed in successive sequential washing steps using: (i) distilled water or 1-5 % NH_3 or 1-5 % Na_2CO_3 solution; (ii) leather/solution ratio 1/10; (iii) 24 + 24 + 24 h for time of contact; and, (iv) holding temperature of 296 ± 5 K with 50 rpm shaking. After each 24 h washing step, an equal volume of fresh solution was added to the residue. Solutions were stored at 277 K and further characterized regarding pH, chromium (VI), total chromium and total organic carbon (TOC). The final residue was dried at 373 ± 2 K. Concentrated sulfuric acid, ammonia, sodium carbonate and other reagents were of analytical grade. The acid extracts, washing solutions and residue materials were characterized according to the standards listed in Table 1 and the results are presented in Table 3.

Table 1. Chemical and physical methods for characterization of the waste material, acid extracts, washing solutions and developed composites

Parameter	Method
Chemical characterization of leather material and composites	
pH	ISO 4045:2008
Chromium (VI)	ISO 17075:2007
Amines (azo colorants)	ISO/TS 17234:2003
Phenols	ISO 17070:2006
Formaldehyde	ISO 17226-2:2008
Cadmium and Lead	Acid digestion & US EPA 7000B:1998
Total organic carbon (TOC)	EN 13137:2001
Total Cr	Acid digestion & US EPA 7190B:1986
Eluate from leaching test of leather waste and composites by EN 12457-2:2002 (ratio 1:10)	
pH Analytics	ISO 4045:2008
Chromium (VI)	ISO 17075:2007
Total Cr	US EPA 7190B:1986
Cadmium, mg/kg	
Plumb, mg/kg	
Physical characterization of developed composites	
Density	ISO 2781:2008
Hardness	ISO 868:2003 - Shore A
Abrasion resistance	EN 12770:1999
Tear strength	EN 12771:1999
Tensile strength and Elongation at break	EN 12803:2000
Water absorption and desorption	EN ISO 20344:2004
Fatigue (Ross Flex resistance)	BS 5131-2.1:1991

Biodegradability Tests

Biodegradability tests were performed in the acid extracts obtained at 313 K according to ASTM D5210-92(2007) [54] standard, using sludge from leather factory "Curtumes Aveneda", anaerobic Wastewater Treatment Plant (WTP) and beer producer "Unicer", anaerobic WTP. Measurement of biogas and calculations were performed by using the WTW OxiTop® Control measuring system according to Süßmuth et al [55]. In this method, the total degradation coefficient is calculated according to:

$$D_t = [(n_{CO2,g;CH4,g} + n_{CO2,l})/n_{C,theo}] \times 100\% \qquad (1)$$

Where:
D_t – coefficient of total biological degradation, in percentage;
$n_{CO2,g;CH4,g}$ – number of moles of carbon dioxide and methane gases formed;
$n_{CO2,l}$ – number of moles of carbon from carbon dioxide formed in the aqueous phase
$n_{C,theo}$ – theoretical number of moles of carbon in the test solution or material.

Rubber Compounding Tests

The aim of these tests was assessing the use of the de-chromed and washed/neutralized residues, as reinforcing additives or charges in rubber composites usable as soles and its non-hazardousness at the end of their life cycle, closing the leather cycle. Thus, the original leather waste and residues obtained after treatment at 313 K (72 + 72 h) and washed/neutralized with H_2O, NH_3 or Na_2CO_3 without shaking, were shredded to approximately ≤1mm in order to obtain leather short fibers and granules. The NBR material was from Kumho Petrochemical®, medium viscosity (45-50 Mooney), grade NBR 35 LM. Other chemicals used are listed in Table 4 and were of rubber grade. All the materials were conditioned in standard laboratory atmosphere (296 ± 2 K and 50 ± 5% relative humidity) for composites preparation.

The NBR rubber compound was prepared by following the recipe listed in Table 4 and using an industrial Bambury® mixer. To avoid industrial mixer contamination, compounding of NBR with the leather materials was made on a laboratory-scale open two-roll rubber mixing mill from Roca & Guix®.

Samples with 15 and 25 phr of leather material to the NBR rubber compound were prepared. The formulations and references given to the composites are also shown in Table 4. A control mix with no residue was also prepared for comparison and it was referenced as NBR.

The vulcanization time of compounds was determined at 438 K using an oscillating disc Gibitre® Rheometer. Vulcanization was carried out in a lab hydraulic heated press at 438 K for the optimum curing times obtained by rheometry, using a 20 x 20 cm² with 0.2-0.3 cm thickness mould. All the composites plaques were produced in triplicate and stored at 296 ± 2 K. Specimens for characterization were mechanically cut out from the vulcanized plaques. The composites were physically and chemically characterized according to the standards listed in Table 1 and the results are presented in Figure 1.

N.	Equation	Where:
(2)	$\dfrac{\Delta M}{\Delta M_0} - 1 = a_f \dfrac{mf}{mp}$	ΔM is the increase in torque moment of the compound with leather ΔM_0 is the increase in torque moment of the formulation with no leather mf is the mass in the parts by weight of leather mp the mass in parts by weight of polymer
(3)	$n = -\dfrac{\ln(1-v_r) + v_r + Xv_r^2}{V_0\left(v_r^{1/3} - 0.5v_r\right)}$	v_r is the volume fraction of polymer in the swollen mass X is the Flory-Huggins polymer-solvent dimensionless interaction term (0.479 for NBR) V_0 is the molar volume of the solvent (V_0 = Molar mass of solvent/Density of the solvent that for toluene is 106.3 cm³/mole).
(4)	$S_{ratio} = 100\,\dfrac{W_{rs} - W_r}{W_r}$	W_{rs} is the weight of the swollen sample W_r is the rubber sample initial weight

Vulcanization characteristics were determined by following the ASTM D2084-07 standard [56], at approximately 438 K, with a Gibitre® Rheometer, giving data for calculating the minimum torque M_L (dN.m); the increase in torque moment ΔM (dN.m); the optimal vulcanization time t_{90}; and the activity of leather a_f based on equation (2) [32]. These results are presented in Table 5. The number of crosslinks in rubber materials, n (moles cm^{-3}), was determined on the basis of solvent-swelling measurements by following ISO 1817:2005 [57] (toluene at 296 ± 2 K during 24 hours) by application of the Flory-Rhener equation (3) [58]. The swelling ratios S_{ratio} (percentage swelling) of vulcanizates were calculated by using equation (4) [30]. These results are presented in Figure 2.

RESULTS AND DISCUSSION

Sample Characteristics

Table 2 presents the characteristics of the leather used in the de-chroming tests. Samples had about 3.1% of chromium and 53.6% of TOC, and contained Cr(VI), azo-colorants, pentachlorophenol, formaldehyde and metals (cadmium and lead) below the respective threshold values. As the value for total chromium is above 3 g kg^{-1} and below 5% this waste is considered non-inert and non-hazardous according to the Portuguese law [51]. The acid nature of leather was reflected in the pH according ISO 4045:2008 and in the eluate by the EN 12457-2:2002 leaching test. This eluate had <0.01 mg L^{-1} of Cr(VI) and 35 mg L^{-1} of total Cr, thus the leather waste requires pre-treatment for disposal at hazardous landfills (total Cr in eluate exceeding the respective threshold values, respectively 2 and 10 mg L^{-1} according PT [51] and EU [52] legislations).

De-chroming Tests

Table 3a summarizes the results obtained by a 72 + 72 hours periodic process of solution renewal using a 25-mL concentrated H_2SO_4 L^{-1} solution: (i) at 296 K with leather/solution ratio of 1/10; and, (ii) 313 K with leather/solution ratio of 1/5. The first extraction at 296 K and 313 K, recovered about 21% and 40% of the chromium contained in the leather, respectively. The second sequential 72 h extraction recovered about 12% and 22% of the chromium, respectively at 296 K and 313 K. Thus, as expected, the increase in temperature promoted higher Cr removal, and after the 2 steps of extraction about 63% of Cr was leached at 313 K compared to 33% of Cr at room temperature. The leather matrix attack, evaluated through TOC in the solution, was in the range of 3-20%. In consequence, the general leather appearance remained despite some visible shrinkage. Temperature is essential for organic matter attack. Results showed that chromium and organic matter were removed from leather into the solution, step by step, according to a decreasing pattern where the first extraction is the most significant. The Cr concentration in the solutions obtained at 313 K was in the range of 1.5-3 g L^{-1} and gives them some interest as addition to leather retanning baths.

Mass loss and total Cr on the 72 + 72 hours de-chromed residues are in general agreement with the values estimated through the mass balances using TOC and total Cr of the acid extracts. These residues subjected to the EN 12457-2:2002 leaching test released more Cr than the original leather scrap due to acid chromium-collagen bond destabilization and incomplete washing out of de-linked chromium.

Table 2. Characteristics of the leather waste used

Parameter	Result	Parameter	Result
Chemical characterization		*Eluate from leaching test*	
pH	3.44	pH	3.48
Chromium (VI), mg/kg	<3	Cr(VI), mg/L	<0.01
Amines, mg/kg	<30	Total Cr, mg/L	35.0
Phenols (tri, tetra e penta), mg/kg	<5		
Formaldehyde, mg/kg	<10		
Cadmium and lead, mg/kg	<100		
TOC, %	53.6		
Total Cr, %	2.8		
Environmental law specification regarding total Cr and Cr (VI)			
	Portuguese law limit [51]	European Union limit [52]	
Total Cr, g/kg	3 inert	Not applicable	
Total Cr, %	5 non-hazardous	Not applicable	
Chromium (VI) in eluate, mg/L	0.1 non-hazardous	Not applicable	
Total Cr in eluate, mg/L	2 non-hazardous	10 non-hazardous	

Table 3. Chemical characterization of acid extracts, washing solutions and residues

Extraction temperature. Parameter	296 K			313 K	
a. *Acid extracts and residue*	72 h	72 + 72 h	72 h		72 + 72 h
pH	0.88 ± 0.01	0.86 ± 0.01	0.89 ± 0.02		0.87 ± 0.02
Total Cr, mg/L	679 ± 108	469 ± 19	2750 ± 111		1540 ± 33
Cr recovery, %	21.2 ± 3.8	11.8 ± 0.7	40.2 ± 1.9		22.5 ± 0.9
Cr (VI), mg/L	<3	<3	<3		<3
TOC, mg/L	1378 ± 38	570 ± 50	8344 ± 797		4131 ± 478
TOC recovered, %	2.6 ± 0.1	1.1 ± 0.1	15.6 ± 5.5		7.7 ± 0.8
Loss of mass, %	NA	4.7 ± 0.3	NA		25.9 ± 0.5
Cr recovery, %	NA	32.1 ± 0.7	NA		60.1 ± 0.9
Cr in residue eluate, mg/L	NA	101.9 ± 10.7	NA		266.2 ± 21.7
b. *Washing extracts and residue*					
Water	*With shaking* 24 h; 48 h; 72 h		*With shaking* 24 h; 48 h; 72 h		*Without shaking* 24 h; 48 h; 72 h
Total Cr, mg/L	35 ± 3; 12 ± 1; 7 ± 1		34 ± 1; 10 ± 1; 2 ± 1		33 ± 1; 7 ± 1; 2 ± 1
Cr recovery, %	~1.5; 0.5; 0.3		~1.5; 0.5; 0.1		~1.5; 0.4; 0.1
Cr in residue eluate, mg/L	NA; NA; 34.0 ± 0.8		NA; NA; 4.8 ± 0.1		NA; NA; 9.0 ± 0.1
NH$_3$ washing solution					
Total Cr, mg/L	67 ± 3; 11 ± 1; 11 ± 1		53 ± 1; 11 ± 2; 3 ± 1		47.3 ± 1; 13 ± 2; 5 ± 1
Cr recovery, %	~2.5; 0.5; 0.5		~2; 0.5; 0.2		~2; 0.5; 0.25
Cr in residue eluate, mg/L . Na$_2$CO$_3$ washing	NA; NA; 17.1 ± 0.1		NA; NA; 53.2 ± 0.9		NA; NA; 34.0 ± 0.6
Total Cr, mg/L	40 ± 2; 10 ± 1; 10 ± 1		25 ± 5; 16 ± 3; 3 ± 2		25 ± 5; 7 ± 1; 5 ± 1
Cr recovery, %	~1.75; 0.5; 0.5		~1; 0.75; 0.2		~1; 0.4; 0.25
Cr in residue eluate, mg/L	NA; NA; 265.8 ± 20.1		NA; NA; 105.1 ± 10.5		NA; NA; 114.0 ± 10.9

Legend: NA – Not analyzed.

The results of sequentially washing the 72 + 72 hours extracted leathers at 296 K and 313 K with distilled H_2O, NH_3 and Na_2CO_3 solutions and 50 rpm shaking are depicted in Table 3b. The results of washing the 72 + 72 hours and 313 K extracted leathers with the same solutions but without agitation are also included there. In the first washing step with contact time of 24 h, 5% NH_3 and Na_2CO_3 solutions were used. In the second and third washing steps with contact time of 24 h, 1% NH_3 and Na_2CO_3 solutions were used. Results are globally in agreement, indicating the first washing steps as removing higher amount of chromium. Globally chromium recovery is in the 1.65 to 2.75 % range. Although the first objective with the washing step was to render the acid residues less hazardous, in the case of using alkaline solutions, additional objectives could be obtained: (i) making residues more compatible with rubber as indicated by Ravichandran and Natchimuthu [29-31]; and, (ii) to obtain solutions that could be used to neutralize and clean the acid extract, thus contributing to close the chromium cycle. Results indicated that water was the best washing agent, especially for the 313 K acid residue, since it originates a less hazardous ultimate residue (still hazardous according PT regulation and non-hazardous according to EU law).

Biodegradability Tests

Biodegradability tests were performed in triplicate on the acid extracts resulting from de-chroming treatment at 313 K, using sludge samples from two different anaerobic WTP, respectively a leather and a beer factory. Under these test conditions the coefficient of total biological degradation, D_t, as defined in expression (1), for the used standard cellulose material (reference Avicel from Fluka) was 60 ± 10% in 90 days. This coefficient, D_t, for the acid extract gave 35 ± 5% and 30 ± 5%, respectively using the leather and beer factories sludge's. These anaerobic biodegradability tests yielded results with some dispersion depending on sludge concentration and sludge origin.

Rubber Composites Tests

The composites prepared according those conditions indicated in Table 4 showed the physical properties indicated in Figure 1. Density values were determined according to the ISO 2781:2008, Method B, and showed that the addition of leather materials has a negligible effect in the NBR base formulations density (Figure 1a). The density of all NBR composites was in the range of 1.12 ± 0.03 Mg m^{-3}.

Abrasion resistance decreased with leather addition (Figure 1b). Nevertheless, all composites presented abrasion resistance adequate for application in soles of demanding use footwear (<150 mm^3).

With leather incorporation of 15 and 25 phr, tear resistance increased relatively to the base NBR formulation (Figure 1c) by: (i) 56 and 90% with C1 (leather waste fibers without treatment) incorporation; (ii) 27 and 26% with C2 (leather waste fibers after acid extraction and washed with H$_2$O without shaking) incorporation; (iii) 13 and 22% with C3 (leather waste fibers after acid extraction and washed with NH$_3$ without shaking) incorporation; and, (iv) 6 and 7% with C4 (leather waste fibers after acid extraction and washed with Na$_2$CO$_3$ without shaking) incorporation. All composites are adequate for applications in soles for normal and demanding footwear, whose minimum values of tear resistance are respectively 7 and 10 N mm^{-1}. The considerable increase in tear strength up to specified limits when non-treated leather residue additive was added is due to the fibrous nature of leather that effectively prevents the growth of the test specimen crack. After extraction and washing, the leather material looses some of its fiber structure resulting in lower reinforcement. Thus, as previously found [59], the above-referred increase on tear resistance seems to be essentially a physical phenomenon.

Tensile strength decreased relevantly with leather additives (Figure 1d) and none of the composites fulfills the minimum tensile strength specification of 8 MPa. Addition of leather materials had a negligible effect in the elongation at break of NBR base formulations (Figure 1e). The elongation at break of all composites was in the range $550 \pm 15\%$, satisfactory for soles application.

According BS 5131/2.1:1991 standard, all NBR composites till 25 phr incorporation gave around 0.01 mm kcycle^{-1}, and they are satisfactory with respect to flexure resistance.

The decrease on abrasion resistance and tensile strength with the additives incorporation may partially be explained by their size and the consequent voids that are created in the matrix by the addition. Upon abrasion and tension the voids grow in size and interact between them, leading to material de-bonding and failure. As indicated by Ravichandran and Natchimuthu [30,31], in the case of non-treated leather fibers, a further negative contribution comes from the acidic and hydrophilic leather nature, leading to poor adhesion between the leather and the matrix. Even though the affinity between leather materials and elastomeric matrix was improved by treating the material with

NH_3 or Na_2CO_3, no sensible benefits were promoted by these treatments in the present study.

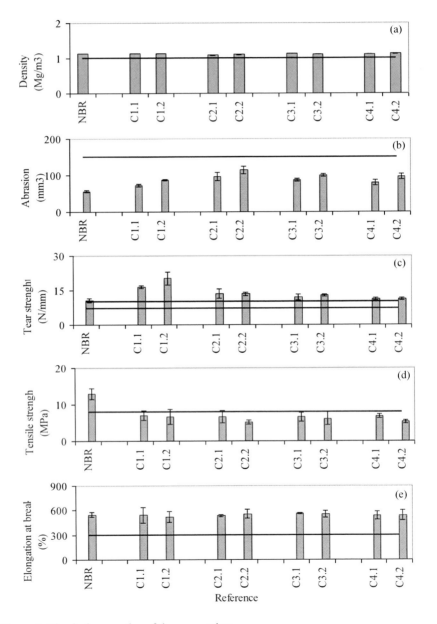

Figure 1. Physical properties of the composites.

Table 4. Rubber compounds and composites formulations

NBR Formulation components	phr	Leather additive	Composites
Acrylonitrile butadiene rubber	100		15 phr 25 phr
Filler (Silica VN3)	30		
Activator (Polyethylene glycol, Stearic acid, Zinc carbonate)	6.0 ± 1.0	Initial leather waste	C1.1 C1.2
Plasticizer (DIBP)	10.0 ± 1.0	Materials after extraction at	
Antioxidant (Lovinox® 22 M46, Vulcanox® MB2, Antiozonant AC052)	3.5 ± 0.5	313 K and static washing with:	
Accelerator (Dibenzothiazyl disulphide - MBTS, Tetra methyl thiuram disulphide - TMTD)	1.5 ± 0.5	. Distilled water . Na$_2$CO$_3$. NH$_3$	C2.1 C2.2 C3.1 C3.2 C4.1 C4.2
Vulcanization agent (Sulphur SU95)	1.5		

phr - parts per hundred of rubber; g of component per 100 g of rubber polymer

The base NBR vulcanizate as well as the obtained composites were chemically characterized by following those procedures listed in Table 1. All samples gave hexavalent chromium, total chromium, cadmium and lead below the threshold values. The leaching test of these materials were determined by following EN 12457-2:2002 protocol and gave eluates whose hexavalent and total chromium were below the PT and EU threshold values, confirming them as non-hazardous wastes at the end of life.

The rheometric properties of composites produced are in general agreement, as observed in Table 5. From the rheometric parameters obtained, it can be concluded that the addition of leather materials to NBR till 25 phr: i) increases the viscosity of the mix (M_L), which is partially attributed to the fibrous structure of leather materials and its more rigid nature; ii) has no effect in the torque ΔM or sligthly decreases it, suggesting no influence of the additives in the crosslinking degree of vulcanizates [32,33]; iii) slightly decreases vulcanization time, t_{90}, particularly for C3 and C4 composites, suggesting that alkaline additives shorten the vulcanization reaction of the elastomers; and, iv) develops no additive activity (a_f).

Table 5. Rheometric properties of the composites

Reference	M_L (dN.m)	ΔM(dN.m)	t 90(min)	a_f
NBR	14.9	43.9	3:46	-
C1.1	16.8	41.6	3:45	<1
C1.2	18.3	37.5	3:34	<1
C2.1	16.4	39.5	3:45	<1
C2.2	16.8	32.8	3:46	<1
C3.1	14.7	31.5	3:01	<1
C3.2	16.2	35.2	2:56	<1
C4.1	16.6	43.3	3:00	<1
C4.2	17.9	33.7	2:55	<1

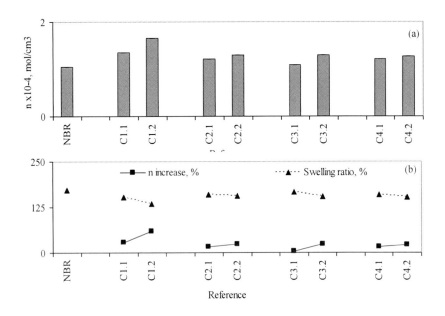

Figure 2. Composites rheometric properties

The crosslink density values (n) of the vulcanizates and composites measured in toluene are presented in Figure 2a. The crosslink density increase (n increase) and swelling ratio, expressed as the variation in weight %, are presented in Figure 2b. Based on equilibrium swelling in toluene, it was observed that the incorporation of non-treated leather brought an increase in the crosslink density of the vulcanizates, which could indicate that leather takes part in the crosslinking of vulcanizates as observed with other systems [32,33]. The treated materials caused a very small increase in the crosslink density of the vulcanizates. When leather-based additives were added, the swelling ratio values in toluene were slightly reduced, particularly in the case of the non-treated material. This can be due to some physical constraining role of leather at the elastomeric matrix in the organic solvent. Thus, swelling tests also suggest that there is some reinforcing action connected with the non-treated leather based additive.

CONCLUSIONS

The present work presents the basis for developing a low cost treatment route for recycling finished leather scrap. The approach is to recover chromium by holding the leather scrap in a 25 mL static H_2SO_4 L^{-1} solution under a liquid/leather ratio of 10 (w/w) for 6 days at 313 K followed by solid separation and washing the residue. The obtained acid chromium extracts could preferably be used as additives to chromium sulfate tanning solutions in areas where leather tanning plants operate. As these extracts present reasonable anaerobic biodegradability around 25-40 %, other applications and treatment alternatives are possible. The non-treated leather scrap and de-chromed and water washed residues after grinded to ≤ 1 mm may be added till 25 phr to NBR composites for production of shoe rubber soles with improved tear resistance and a slight decrease on tensile strength. The addition of leather scrap fibers without treatment give the best performing composites. Under an environmental point of view, when de-chromed leather scrap after washing with water is added to NBR elastomer up to 25 phr, the end-products obtained may be considered non-hazardous wastes at their end-of-life. This is an interesting achievement since chromium in eluate from chromium tanned leather scrap leaching tests is often above the threshold value for admission of wastes in non-hazardous landfills.

ACKNOWLEDGMENTS

The authors acknowledge the financial support of "Fundação para a Ciência e Tecnologia", under the research project POCI/AMB/62704/2004, the leather company "Curtumes Aveneda" and the beer company "Unicer" for supplying sludge's for biodegradability tests, and the rubber production company Procalçado - Produtora de Componentes para Calçado, S.A., particularly Sr. Pinto, Dr. Fátima Pinto, Dr. José Pinto and Eng.º Rui Russo for facilitating the use of lab, scale-up and industrial materials and equipments for making the rubber composites tests.

REFERENCES

[1] Gish AJ. *J. Am. Leather Chem. Ass.*, 1999;94:43-47.
[2] Markus RT, Duman A, Wyler A. *J. Soc. Leather Technol. Chem.*, 1990;74:74-77.
[3] Manzo G, Fedele G. *Cuoio Pelli Mat. Concianti*, 1992;68:491-501.
[4] Manzo G, Fedele G. *Cuoio, Pelli, Mat. Concianti*, 1989;65:45-51.
[5] Zhang M, Shi B. *J. Soc. Leather Technol. Chem.*, 2004;88:236-241.
[6] Przepiorkowska A, Janowska G. *J. Soc. Leather Technol. Chem.*, 2003;87:223-226.
[7] Addy V. *Leather International*, 2004;5:44-48.
[8] Dogru M, Midilli A, Akay G. *Energy Sources*, 2004;26:35-44.
[9] Midilli A, Dogru M, Akay G. *Energy Sources*, 2004;26:45-53.
[10] Bowden W. *J. Am. Leather Chem. Ass.*, 2003;98:19-26.
[11] Bahillo A, Armesto L, Cabanillas A. *Waste Management*, 2004;24:935-944.
[12] http://www.apiccaps.pt/. Acessed 2 June 2009.
[13] Almeida MF, Ferreira MJ. *Environmental Geotechnics*, Seco e Pinto, Ed. Balkema Rotterdam (1998).
[14] Ferreira MJ, Almeida, MF, Pinto T. *J. Soc. Leather Technol. Chem.*, 1999;83:135-139.
[15] Okamura H, Shiray K. *J. Am. Leather Chem. Ass.*, 1976;71:173-179.
[16] Ferreira MJ, Almeida MF, Pinho SC, Santos IC. *Waste Management*, 2009;doi:10.1016/j.wasman.2009.12.006.
[17] Manzo G, Fedele G. *Cuoio, Pelli, Mat. Concianti*, 1980;56:743-758.
[18] Wojciech L, Miesczyslzw G, Ursula M. *PCT Int. Appl.* WO. 9803685 (1998).
[19] Tomaselli M, Girardi V, Liccardi, C. *Cuoio, Pelli, Mat. Concianti*, 2000;76:167-175.
[20] Manzo G, Fedele G, Colurcio A. *Cuoio, Pelli, Mat. Concianti*, 1993;69:63-69.
[21] Manzo G, Fedele G. *Cuoio, Pelli, Mat. Concianti*, 1994;70:17-20.

[22] Muñoz J, Maldonado M, Rangel A. *J. Am. Leather Chem. Ass.*, 2002;97:83-88.
[23] Jianzhong M, Lingyun L. *Leather Technol. Chem.*, 2004;88:1-5.
[24] Kasparkova V, Kolomaznik K, Burketova L, Sasek V, Simek, L. *J. Am. Leather Chem. Ass.*, 2009;104:46-51.
[25] Cabeza LF, Taylor MM, *J. Am. Leather Chem. Ass.*, 1998;93:83-98.
[26] Cabeza LF, Malcon AJ. *J. Am. Leather Chem. Ass.*,1998;93:299-315.
[27] Ferreira MJ, Xará E, Almeida MF, Barla M, Ferrer J. *J. Soc. Leather Technol. Chem.*, 2000;84:271-275.
[28] Ferreira MJ, Xará E, Almeida MF, Barla M, Ferrer J. *J. Soc. Leather Technol. Chem.*, 2001;85:193-198.
[29] Ravichandran K, Natchimuthu, N. *Polímeros: Ciência e Tecnologia*, 2005;15:102-108.
[30] Ravichandran K, Natchimuthu N. *Polymer Int.*, 2005;54:553-559.
[31] Natchimuthu N, Radhakrishnan G, Palanivel K, Ramamurthy K, Anand JS. *Polymer Int.*, 1994;33:329-333.
[32] Przepiorkowska A, Chronska K, Zaborski M. *J. Hazard. Mater.*, 2007;141:252-257.
[33] Chronska K, Przepiorkowska A. *J. Hazardous Mater.*, 2008;151:348-355.
[34] ISO 4045:2008. *Leather.* Determination of pH.
[35] ISO 17075:2007. *Leather. Chemical tests.* Determination of chromium VI content.
[36] ISO/TS 17234:2003. *Leather.* Chemical tests. Determination of certain azo colorants in dyed leathers.
[37] ISO 17070:2006. *Leather.* Chemical tests. Determination of the content of pentachlorophenol in leather.
[38] DIS 17226-2:2008. *Leather.* Chemical tests. Part 2: Determination of formaldehyde content in leather by colorimetric analysis.
[39] US EPA 7000B:1998. Flame Atomic Absorption Spectroscopy (Consolidate FLAA).
[40] EN13137:2001. *Characterisation of waste.* Determination of total organic carbon (TOC) in waste, sludges and sediments.
[41] ISO/DIS 5398-1:2005. *Leather.* Chemical determination of chromic oxide content. Part 1: Quantification by titration.
[42] EN 12457-2 – *Characterization of Waste Leaching* – Compliance Test for Leaching of Granular Waste Materials and Sludges – Part 2: One stage batch test at a liquid to solid ratio of 10 L/kg with particle size below 4 mm, CEN/TC 292/WG 2, Brussels.

[43] US EPA 7190:1986. *United States Environmental Protection Agency (US EPA)*. Chromium – AA, Direct Aspiration.
[44] ISO 2781:2008. *Rubber, vulcanized or thermoplastic*. Determination of density.
[45] ISO 868:2003. *Plastics and ebonite*. Determination of indentation hardness by means of a durometer (Shore hardness).
[46] Footwear. *Test methods for outsoles*. Abrasion resistance.
[47] EN 12771:1999 (Ed. 1). *Footwear*. Test methods for outsoles. Tear strength.
[48] EN 12803:2000 (Ed. 1). *Footwear. Test methods for outsoles*. Tensile strength and elongation at break.
[49] EN ISO 20344:2004. *Personnel protective equipments*. Footwear test methods..
[50] BS 5131-2.1:1991. *Methods of test for footwear and footwear materials*. Solings. Ross flexing method for cut growth resistance of soling materials.
[51] Decreto-Lei n.º 152/2002, 2002, Diário da República, 23 Maio 2002 (núm.119), *Série I -Ministério do Ambiente e do Ordenamento do Território*.
[52] Council Decision 2003/33/CE of 19 December 2002 establishing criteria and procedures for the acceptance of waste at landfills pursuant to Article 16 and Annex II to Directive 1999/31/EC, *Official Journal of the European Communities*, L 11/27, 16.1.2003.
[53] DIN 38414-S4:1984. German standard methods for the examination of water, waste water and sludge. Group S (sludge and sediments). *Determination of leachability by water* (S4).
[54] ASTM D5210-92(2007). *Standard test method for determining anaerobic biodegradability of plastic materials in the presence of municipal sewage sludge*.
[55] Süßmuth R, Doser C, Lueders T. Application report 0600421e. *Universitat Hohenheim*, Germany, (1999).
[56] ASTM D 2084-07. *Standard test method for rubber property*. Vulcanization using oscillating disk cure meter.
[57] ISO 1817:2005 *Rubber, vulcanized*. Determination of the effect of liquids.
[58] Flory PJ. *Principles of polymer chemistry*, Ithaca, N.Y.: Cornell University (1953).
[59] Ferreira MJ, Almeida MF, Freitas, F. *Polym. Eng. Sci.*, 2009;DOI 10.1002/pen.21643.

In: Biodegradable Polymers ...
Editors: A. Jimenez and G. E. Zaikov

ISBN 978-1-61209-520-2
© 2011 Nova Science Publishers, Inc.

Chapter 15

HOW TO SHIFT TOUGHNESS OF PLA INTO NON-BREAK AREA AND TO CREATE HIGH IMPACT FLAX FIBRE REINFORCEMENTS

R. Forstne[1,*] and W. Stadlbauer
Transfercenter für Kunststofftechnik GmbH, Franz-Fritsch-Straße 11, A-4600 Wels, Austria

ABSTRACT

PLA compounds with various impact modifiers were tested and improvements in Charpy impact strength up to factors of ten were achieved with new Core-Shell type modifiers opening the non-break area for PLA. Reinforcements with flax fibres increased tensile properties and dimensional stability on costs of toughness.

Keywords: impact modification, Young's modulus

INTRODUCTION

In recent years poly(lactic acid), PLA, as a green polymer has attracted scientific and industrial attention since it can be processed similarly to

---

[*] Tel.:+43 7242 2008 1022, e-mail: reinhard.forstner@tckt.at

polyolefins and it is known to be biodegradable. Its good mechanical properties e.g. high stiffness and high tensile strength combined with good optical properties (transparency) made PLA successful in many packing and household applications, but due to the low thermal dimensional stability, big markets are still prohibited. In order to make PLA e.g. feasible for automotive industry or other high performance markets, new composites need to be developed to overcome the low thermal dimensional stability and to improve toughness. In the past, different attempts with reinforcements like jute [1], microcrystalline cellulose (MCC) [2] or flax [3] were made to improve tensile properties in general on costs of impact strength. Recent studies on impact modification of PLA already showed an increase of toughness by using natural rubber [4,5] or commercial petrol based impact modifiers [6]. The goal of this study was to improve impact properties with different commercial available additives as well to improve thermal stability and stiffness of PLA by incorporating flax fibers.

EXPERIMENTAL

Materials

A commercial Nature Works PLA grade with an MFI of 2-4 g $10min^{-1}$ (190° C, 2.16kg) was used as matrix material in this study. The three different grades of impact modifiers were supplied by Arkema and DuPont.

Processing

The compounds were processed in a parallel twin screw extruder (Prism TSE 24 HC) with a screw length of 28D and gravimetric side feeder system for adding the additives and fibers, which were pelletized in a Kahl laboratory before compounding. The neat PLA grade was dried for eight hours in a Motan Luxor 120 at a temperature of 60 °C and the Flax fibers were dried in an oven before compounding for eight hours at a temperature of 80 °C.

Sample Testing

Dog-bone shaped specimens for mechanical tests were moulded in an injection moulding machine (Engel ES 80) and used in tensile tests. Samples for Charpy Impact tests were cut from these dog-bone specimens into rectangular shape of 80 x 10 x 4 mm³ with the aid of a custom made Haidlmair pneumatic sample chopper.

Tensile tests were performed in a universal testing machine (Zwick TC-FR020 TH) at a crosshead speed of 5mm min^{-1} and the unnotched Charpy Impact strength was determined in a Charpy Impact tester Zwick 5113.300 according to the ISO standards ISO-527-2 and ISO-179 respectively.

The heat deflection temperature (HDT) measurements were performed in a Coesfeld Vicat/HDT tester according to DIN-EN-ISO-75-2 on rectangular shaped specimens of the dimensions 80 x 10 x 4 mm³ with a fiber stress of 1.82MPa (HDT-A) and a heating rate of 2 °C min^{-1}.

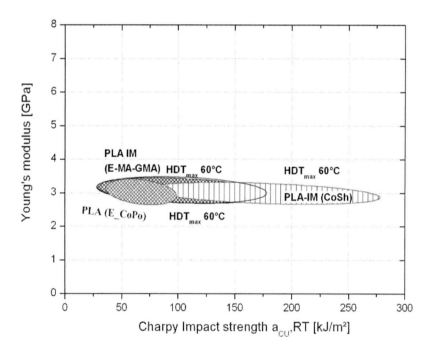

Figure 1. Overview of the mechanical and thermal properties of impact modified PLA.

Optical Microscopy

The morphology of materials was investigated on an Olympus BX 61 optical microscope with a ColorView II camera system pro including evaluation software.

Scanning Electron Microscopy

The composite microstructure was investigated at Linz University using a Zeiss 1540XB scanning electron microscope and at the University of Applied Sciences in Wels with an Tescan Vega II LMU scanning electron microscope.

RESULTS

Impact Modification

PLA compounds of commercial petrol-based impact modifiers, such as E-MA-GMA (ethylene-maleic acid-glycidyl methacrylate) or E-CoPo (Ethylen CoPolymer), as well as core-shell types with contents of 5% to 30% were obtained in a parallel twin-screw extruder and mechanically tested according to ISO standards for composites and plastics. The quality of the individual impact modifiers for PLA can be clearly observed in Figure 1 and Table 1. The addition of E-CoPo resulted in an increase of toughness of more than a factor of three to five followed by the E-MA-GMA types with an increase of a factor up to eight. But both were outperformed by the new core shell types, which improved the impact properties of PLA by a factor up to more than ten.

Naturally, readers will have questions about the absolute values gained for the impact properties of the core shell modified PLA, but the answer lies in the limits of the testing and evaluation method itself. First, one has to be aware of the fact that the samples containing more than 10 wt% of additive were tested in the non-break area, which means that the energy of the pendulum was not completely transferred to the sample resulting in some energy loss to the system. Secondly, the values for the Charpy impact strength were evaluated by a standard software routine, which was delivered with the impact tester and not corrected for the energy loss. However, the most significant result of this series was that the PLA grade was shifted into the non-break area already by the use of industrial amounts of additive (10 wt%).

Table 1. Mechanical data of the impact modified compounds

				Young's Modulus [GPa]	s [GPa]	Charpy impact strength [kJ m^{-2}]	S [kJ m^{-2}]
100 %	PLA	---	---	3.56	0.02	17	0.3
95 %	PLA	5 %	E-CoPo	3.20	0.03	22	3.3
90 %	PLA	10 %	E-CoPo	3.00	0.01	29	5.8
85 %	PLA	15 %	E-CoPo	2.67	0.01	43	3.2
80 %	PLA	20 %	E-CoPo	2.40	0.01	45	2.8
75 %	PLA	25 %	E-CoPo	2.12	0.01	48	8.8
70 %	PLA	30 %	E-CoPo	1.89	0.01	66	11.2
95 %	PLA	5 %	E-MA-GMA	3.20	0.01	24	3.6
90 %	PLA	10 %	E-MA-GMA	2.99	0.02	34	2.2
85 %	PLA	15 %	E-MA-GMA	2.72	0.01	52	3.3
80 %	PLA	20 %	E-MA-GMA	2.41	0.01	85	6.6
75 %	PLA	25 %	E-MA-GMA	2.18	0.03	142	13.6
70 %	PLA	30 %	E-MA-GMA	1.89	0.01	164	4.3
95 %	PLA	5 %	CoSh	3.18	0.02	71	18.5
90 %	PLA	10 %	CoSh	2.85	0.02	273	12.6
85 %	PLA	15 %	CoSh	2.55	0.03	272	1.6
80 %	PLA	20 %	CoSh	2.24	0.01	252	1.7
75 %	PLA	25 %	CoSh	1.97	0.02	233	2.4
70 %	PLA	30 %	CoSh	1.74	0.02	219	2.0

Additional HDT-A tests for annealed samples (1h at 100 °C) showed that the impact modifiers had no nucleating effect in PLA (Figure 1).

Flax Fibre Reinforcement

Natural fibres such as hemp, flax, jute, etc. are generally known to have nucleating effect in semi-crystalline polymers. In order to improve tensile and HDT-properties additional PLA compounds incorporating chopped flax fibers were formulated. The flax fibers had an initial length of 4mm and were pelletized in a Kahl laboratory pelletizer in order to feed the fibers into the compounder.

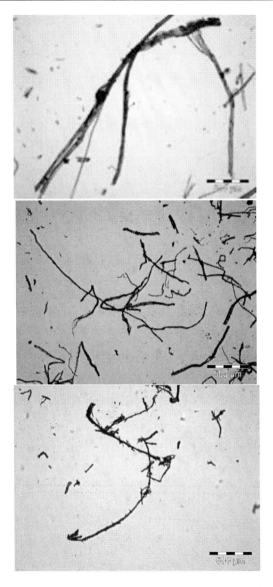

Figure 2. Degradation of Flax fiber length during processing: Optical micrographs of pelletized flax fibers (top-left), flax fibers after compounding (top-right) and flax fibers after injection moulding (bottom left).

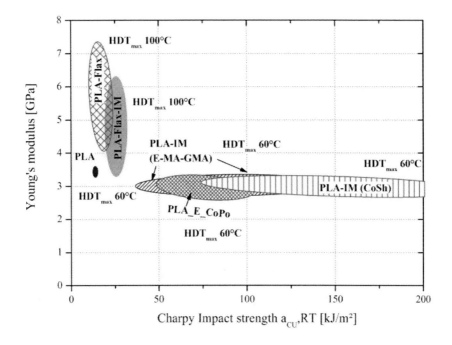

Figure 3. Overview of the mechanical and thermal properties of fibre reinforced and impact modified PLA.

The fiber degradation along the individual processing steps was qualitatively evaluated by optical microscopy (Figure 2). In the case of the compounds, fibers were extracted in a Soxhlet extractor for 4.5 h using CCl_4 as a solvent. As can be seen in Figure 2 a, degradation of fibers during processing was visible but still longer fibers in the range of 1-2 mm (initial length 4 mm) could be found in the compounds or even after injection moulding, where the fiber lengths were about 1-1.5 mm, which guarantee the reinforcement of PLA.

In general, the flax fibers reduce the impact properties compared to the modified PLA compounds but they are still higher than those of the unmodified PLA flax composites or the neat PLA. The E-MA-GMA impact modifier is most effective for PLA-flax fiber compounds, because it increased toughness, tensile and HDT properties compared to the neat PLA-flax fiber compounds, while the other impact modifiers hardly showed any improvement in impact modification (Figure3, Table 2).

Table 2. Mechanical data of the flax fibre and impact modified flax fibre compounds

					Young's Modulus	s	Charpy impact strength	S
					[GPa]	[GPa]	[kJ/m^2]	[kJ/m^2]
100 % PLA	---	---	---	---	3.58	24.0	17	0.3
90 % PLA	10 % Flax	---	---		4.52	23.0	15	1.7
80 % PLA	20 % Flax	---	---		5.87	63.0	15	2.3
70 % PLA	30 % Flax	---	---		7.09	38.0	13	3.1
90 % PLA	---	---	10 % E-CoPo		2.99	11.1	29	5.8
80 % PLA	10 % Flax	10 % E-CoPo			3.65	31.1	21	2.3
70 % PLA	20 % Flax	10 % E-CoPo			4.48	84.8	21	1.3
60 % PLA	30 % Flax	10 % E-CoPo			5.25	88.2	20	2.5
90 % PLA	---	---	10 % E-MA-GMA		2.99	23.3	34	2.2
80 % PLA	10 % Flax	10 % E-MA-GMA			3.81	18.0	21	3.3
70 % PLA	20 % Flax	10 % E-MA-GMA			4.45	9.7	23	2.4
60 % PLA	30 % Flax	10 % E-MA-GMA			5.58	63.5	29	2.4
90 % PLA	---	---	10 % CoSh		3.46	28.0	273	12.6
80 % PLA	10 % Flax	10 % CoSh			3.55	26.5	22	2.4
70 % PLA	20 % Flax	10 % CoSh			4.35	24.9	22	2.2
60 % PLA	30 % Flax	10 % CoSh			5.35	60.9	21	1.6

The maximum dimensional stability for PLA flax compounds was achieved after an annealing step of one hour at 100 °C. The annealed test specimen reached HDT-A values up to 100 °C, which is respectably high for PLA composite materials.

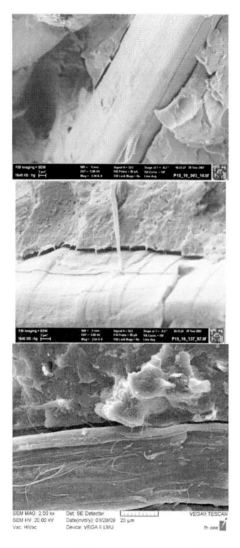

Figure 4. SEM micrographs of a PLA-Flax compound (top-left), a PLA-IM-Flax compound (top-right) and a PLA-Core-shell-Flax compound (bottom left).

Additional SEM measurements (Figure 4) performed on these compounds revealed that some of the used impact modifiers also acted as coupling agents to the natural fibers (flax), which explains the better mechanical performance compared to the non-modified PLA-Flax compounds (Figure 3). A comparison of SEM pictures between PLA-Flax, PLA-IM (E-MA-GMA)-Flax and PLA-IM-Flax (Core-shell) compounds revealed that PLA-IM (E-MA-GMA)-Flax compounds showed good fiber matrix adhesion indicated by the "strings" connecting the fibers with the surrounding matrix, which resulted in a better performance in impact behavior. Due to this fact, additional energy can be consumed by fiber pull out processes compared to the neat PLA-Flax compound. The core-shell impact modified PLA-flax compounds showed a good matrix modification, but poor fiber-matrix adhesion, which in the end resulted in impact properties comparable to the neat PLA-Flax compounds.

CONCLUSION

The impact properties of poly (lactic acid), PLA, with various impact modifiers increased with the increasing amount of impact modifier. A ranking of the impact modifier quality gives best performance for the new core-shell types followed by E-CoPo types using industrial amounts of additive. Flax fibers in PLA-IM compounds decreased the impact properties due to the poor fiber performance or poor fiber-matrix adhesion. Only in the case of E-MA-GMA types, a reasonable increase in impact properties was found since they act both as impact modifiers for PLA (matrix modifier) and coupling agent for Flax fibers.

ACKNOWLEDGMENTS

The authors thank the European Project (IP 515769-2 BioComp) for the financial support, Linz University and the applied University in Wels for performing the SEM measurements.

REFERENCES

[1] Plackett D, Andersen TL, Pedersen WB, Nielsen L. *Comp. Sci. Techn.*, 2003;63:1287-1296.
[2] Mathew AP, Oksman KJ, Sain M. *J. Polym. Sci.*, 2004;97:2014-2025.
[3] Oksman K, Skrifvars M, Selin JF. *Comp. Sci. Tech.*, 2003;63:1317-1324.
[4] Jiménez A, Zaikov, GE., *Recent Advances in Research on Biodegradable Polymers and Sustainable Composites,* Vol.2, Nova Science Publishers Inc., 2008.
[5] Forstner R, Auinger T, Ruf R, Stadlbauer W. PPS 24th, 2008, Stockholm, Sweden, Book of Abstracts.
[6] Forstner R, Stadlbauer W. Polychar 17, 2009, Rouen, France, *Book of Abstracts.*

INDEX

A

Abraham, 164
access, 120, 161, 170
acetic acid, 85, 159, 161
acid, xii, xvi, 3, 21, 24, 38, 51, 52, 53, 63, 138, 155, 159, 161, 182, 185, 188, 208, 209, 210, 212, 213, 215, 216, 217, 218, 219, 221, 223, 230
acidic, 5, 24, 153, 157, 185, 210, 219
acrylonitrile, 210
activated carbon, 209
activation energy, xii, 38, 47, 60, 62, 64
activation parameters, 60, 62
active additives, xiii, 39, 105, 106, 107, 114
active compound, xiv, 105, 107
active site, 8, 120
additives, xiii, xiv, 39, 45, 83, 93, 95, 96, 97, 100, 105, 106, 107, 109, 110, 111, 112, 113, 114, 121, 210, 213, 219, 221, 223, 228
adhesion, 21, 23, 59, 77, 219, 236
adhesions, 20
adhesive interaction, 23
adhesives, 78
adsorption, 8, 23, 121, 208
aesthetic, 208
aesthetics, 211
AFM, 162
agar, 97, 159, 161, 162

agriculture, xv, 151, 153, 155
algae, 96, 152
alkaline media, 5, 17
alkenes, 178
almonds, xv, 137, 138, 139
aluminium, 87, 140, 183
amine, 54, 188, 190
amino, 153, 157, 182, 185, 188, 189
amino acid, 182, 185, 189
amino acids, 182, 185
amino groups, 153, 157
ammonia, 23, 99, 210, 212
ammonium, 95, 124, 196
ammonium salts, 124
amputation, 158
analgesic, 27, 28
anchoring, 23, 121
angiogenesis, 20
annealing, 235
ANOVA, 72
anticancer drug, 25
anticoagulant, 21
anti-inflammatory drugs, 25, 27
antioxidant, xi, 37, 38, 39, 42, 43, 44, 45, 47, 48, 106, 107, 111, 114, 138, 155
antitumor, 27, 28, 153
Argentina, 67, 70, 79, 80
arginine, 189
arteries, 29
artery, 19, 29
arthritis, 29

Index

arthropods, 152
articular cartilage, 21
asparagus, 85, 89
aspartate, 120
aspartic acid, 185
assimilation, 77
atmosphere, 40, 41, 45, 46, 47, 54, 87, 108, 110, 124, 125, 133, 140, 211, 213
atmospheric pressure, 70, 108
atomic force, 162
atoms, 84
atrial septal defect, 19
atrium, 12
attachment, 23
Austria, 227
authentication, 147
autooxidation, 47
axons, 20

B

bacteria, 9, 11, 24, 75, 96, 99, 153, 156, 157, 158, 159, 160, 161, 162
bacteriostatic, 153
barium, 101, 196
barriers, 24, 73
base, 24, 25, 158, 218, 219, 221
basement membrane, 20
baths, 216
beer, 156, 213, 218, 223
bending, 185
beneficial effect, xii, 67
benefits, 138, 220
benign, 25
bicarbonate, 210
biochemical processes, 94
biocompatibility, 2, 18, 21, 24, 25, 34, 152
Biocompatibility, 17, 20, 21, 28
bioconversion, 95
biodegradability, xvi, 11, 68, 152, 170, 182, 194, 195, 196, 197, 200, 204, 205, 208, 210, 218, 223, 224, 226
biodegradable materials, xi, 2, 11, 21, 94, 182

biodegradation, xi, xvi, 1, 2, 3, 5, 9, 11, 12, 13, 14, 16, 24, 78, 94, 96, 97, 194, 195, 196, 198, 199, 200, 201, 202, 203, 204, 208
biogas, 205, 213
biological activities, xv, 151
biological activity, 153
biological media, 9
biologically active compounds, 106
biomass, xv, 152, 169, 170, 171, 175, 178, 195
biomass materials, 170, 171
biomaterials, 17
biomedical applications, xiv, 117, 119, 122, 135, 155
biomedical, cosmetics, xv, 151, 153
biomedicine, xi, 1
biomolecules, 24
biopolymer, xv, 3, 152, 182
biopolymers, xi, xvi, 22, 25, 68, 74, 152, 182, 207
biosynthesis, 2
Biosynthetic pathway, xiv, 117
blends, 4
blood, 9, 23, 24, 29
blood plasma, 24
blood vessels, 29
body fluid, 3, 9
body weight, xiii, 83
Boltzmann constant, 60
bonding, xii, 67, 121, 185, 188, 189, 219
bonds, 73, 77, 152, 188, 189, 195
bone, 13, 17, 18, 19, 21, 122, 228
bone form, 19
bone growth, 19
bone marrow, 21
bones, 19
bounds, 56, 58
bowel, 12, 17, 20
brain, 9, 24
Brazil, 93, 96, 97, 101
bromine, 199
building blocks, 52
burn, 156
butadiene, 210, 221

Index

by-products, xv, 77, 152, 169, 170, 182

C

Ca^{2+}, 123
cables, 20
cadmium, 101, 215, 221
calcium, 101, 154, 196
calibration, 85, 126
cancer, 29
Candida antarctica lipase B (CALB), xiv, 118
candidates, 52, 119, 124
capillary, 86, 172
capsule, 14, 15, 18
carbohydrates, 94
carbon, xiii, 94, 95, 96, 97, 98, 195, 196, 199, 202, 205, 212, 213, 225
carbon dioxide, 195, 196, 199, 205, 213
carbonyl groups, 73
carboxyl, 3, 188, 190
cardiovascular disease, 138
cardiovascular risk, 138
cartilage, 17, 18
case study, xv, 152
casting, xv, 3, 14, 69, 73, 181, 182, 183, 184, 186, 188, 190
catalysis, 14
catalyst, 44, 120, 121, 125, 127, 128
catalytic activity, 128
catalytic system, xiv, 118, 134
cationic surfactants, 123
cattle, 196
C-C, 185
cell death, 17
cell line, 22
cell lines, 22
cell metabolism, 24
cellulose, xii, xv, 51, 53, 56, 58, 59, 60, 62, 63, 64, 152, 156, 170, 171, 175, 176, 177, 178, 200, 201, 218
cellulose fibre, xii, 51, 60, 63
ceramic, 121
chain mobility, 189
chain scission, 4, 5

charge density, 161
Charpy impact strength, xvii, 227, 230, 231, 234
chemical, 2, 23, 26, 52, 59, 77, 78, 94, 95, 98, 127, 147, 162, 171, 195, 199, 210, 211
chemical bonds, 195
chemical properties, 26, 98, 210
chemical reactions, 59
chemicals, 52, 53, 118, 208, 213
chemiluminescence, 38
Chemiluminescence (CL), xi, 38
Chicago, 149
China, 70
chitin, 152, 153
chitosan, xv, 151, 152, 153, 154, 155, 156, 157, 158, 159, 160, 161, 162
Chitosan, viii, xv, 151, 152, 153, 154, 155, 158, 159
chlorine, 84, 199
chloroform, 125, 132
cholesterol, 138
chromatograms, 175, 176
chromatography, xv, 126, 169
chromium, xvi, 193, 194, 196, 202, 204, 207, 208, 209, 210, 211, 215, 216, 218, 221, 223, 225
Chromium sulfate, xvi, 207
chronic diseases, 25
chronic irritation, 24
circulation, 71
classification, 138, 139, 147
clean air, 198
cleaning, 77
cleavage, 3, 4, 12, 95
climate, 172
closure, 15, 20
C-N, 185
CO2, 11, 94, 95, 195, 196, 198, 199, 201, 202, 203
coatings, 79, 158, 163, 191
cobalt, 101
collagen, xvi, 18, 19, 22, 23, 194, 196, 200, 201, 202, 203, 208, 209, 216
color, xi, 209

combustion, 208
commercial, xi, xiii, 2, 20, 69, 84, 85, 86, 87, 88, 89, 91, 124, 140, 175, 177, 228, 230
communication, 30
communities, 100
community, 153
compaction, 208
compatibility, 72, 74, 84
compatibilizing agents, 78
compensation, 60
complement, 209
compliance, xiii, 84, 85, 87
composites, xi, 11, 52, 53, 54, 56, 58, 59, 63, 64, 155, 170, 208, 212, 213, 214, 218, 219, 220, 221, 222, 223, 224, 228, 230, 233
composition, xiii, 2, 8, 23, 84, 87, 89, 95, 138, 146, 189, 199, 204
compost, 11, 94, 95, 96, 97, 98, 99, 100, 101
composting, 9, 94, 95, 96, 98, 99, 100, 102, 204, 205
compounds, xv, xvi, xvii, 38, 43, 87, 95, 99, 106, 107, 110, 114, 138, 146, 162, 169, 171, 172, 175, 176, 178, 182, 193, 194, 214, 221, 227, 228, 230, 231, 233, 234, 235, 236
compression, xii, xv, 54, 67, 69, 70, 72, 78, 108, 181, 182, 183, 184, 186, 187, 190
condensation, 53
conductivity, 142
configuration, 39
connective tissue, 18
consensus, 195
conservation, 139
constant rate, 29
consumption, 43, 54, 182
contact time, 218
contamination, 75, 107, 160, 213
control group, 19
controversial, 161
conventional composite, xii, 52
COOH, 189
cooling, 87, 108, 140, 141

cooling process, 141
coordination, 124
copolymers, 18, 25, 32
correlation, 11, 157
cosmetic, 155
cosmetics, xv, 151, 153, 155
cost, 53, 64, 138, 182, 195, 210, 223
covalent bond, 121, 146
covalent bonding, 121
covering, 120
cracks, 56, 72, 96
crop, 139
crops, 138
crystalline, 5, 8, 11, 14, 32, 90, 109, 122
crystallinity, xiii, 2, 5, 8, 11, 14, 15, 68, 78, 109, 112
crystallisation, 141
crystallization, 26, 40, 78, 109, 141, 142, 143, 147
crystallization kinetics, 26
cultivars, xiv, 137, 139, 142, 143, 144, 145, 146, 147
culture, 11, 97, 157, 199
culture medium, 199
cure, 226
cycles, xi, xv, 87, 169, 172, 175, 176, 177
cystine, 185
cytokines, 21
cytoskeleton, 22

D

deacetylation, 152, 157, 159, 163
decomposition, 44, 45, 94, 95, 145
defects, 17, 20, 44
deformation, 64, 74
degradation, xii, xiii, xiv, xvi, 2, 3, 4, 5, 7, 8, 9, 10, 11, 12, 14, 16, 19, 24, 27, 43, 44, 45, 47, 54, 68, 71, 75, 76, 77, 78, 79, 81, 84, 89, 93, 94, 95, 96, 99, 100, 101, 105, 110, 118, 119, 122, 132, 134, 139, 145, 146, 147, 155, 170, 175, 177, 178, 193, 195, 198, 201, 202, 203, 204, 213, 218, 233
degradation mechanism, 4, 14, 75, 81

degradation process, xiii, 2, 14, 16, 19, 44, 45, 68, 75, 76, 79, 99, 100, 132, 175, 178, 202
degradation rate, xvi, 5, 10, 24, 45, 47, 94, 122, 146, 193
degree of crystallinity, 8, 26
dehydration, 134
dehydrochlorination, 84
denitrifying, 11
density values, 222
dental caries, 156
deposits, 18
derivatives, 85, 119, 121, 170
dermatitis, 160
dermis, 199
desorption, 28, 212
detectable, 21, 87, 88, 89
detection, 40, 108
diabetic patients, 158
dialysis, 124
Differential Scanning Calorimetry (DSC), xi, xv, 38, 87, 90, 108, 137, 138, 140
diffusion, 10, 27, 106, 108, 154, 190
diffusion mechanisms, 28
diffusion process, 27
diffusion rates, 106
diffusivity, 28
digestion, 154, 196, 212
discriminant analysis, 140, 144, 146, 147
discrimination, xv, 137
diseases, 28, 118
dispersion, 39, 54, 55, 59, 122, 144, 147, 182, 183, 190, 210, 218
displacement, 141
distilled water, 71, 183, 196, 211
distribution, 107, 152
DNA, 153, 154
dogs, 28
DOI, 92, 226
donors, 29
dosage, 26
dosing, 28
drug delivery, 25, 26, 27, 28, 29, 34, 154, 156, 163
drug release, 2, 25, 26, 27

drugs, 25, 26, 27, 28, 118
dry matter, 71
drying, xv, 71, 72, 181, 182, 183, 184, 189, 196
DSC, xi, xiv, xv, 38, 40, 41, 42, 43, 45, 87, 90, 105, 108, 109, 111, 112, 118, 137, 138, 139, 140, 141, 142, 143, 144, 147
dumping, 194
dyes, 162, 208

E

E.coli, 24, 157
earthworms, 77
ecology, 102
economical cost, 158
effusion, 18
elaboration, xiv, 117, 119, 122, 132, 134, 135, 137
elastic modulus, xiv, 64, 106, 114
elastomers, 221
electron, xv, 6, 8, 94, 97, 113, 169, 171, 173, 174, 230
electron microscopy, xv, 6, 8, 169
elongation, xiv, xv, 64, 74, 106, 181, 184, 189, 190, 219, 226
e-mail, 227
emboli, 28
embolization, 28
emission, 42
emulsions, 154, 163
encapsulation, 25, 154
endothelial cells, 19, 21
endothelium, 19, 21
endothermic, 141, 142
energy, xiii, 39, 60, 93, 94, 95, 141, 170, 208, 230, 236
energy recovery, 170
engineering, 21, 23, 154, 156
entrapment, 26
environment, xi, xiii, 1, 2, 7, 8, 11, 27, 53, 73, 75, 93, 94, 118, 158, 182, 184
environmental conditions, 75, 94
environmental contamination, 95
environmental factors, 172

environmental impact, xvi, 194, 195, 208
Environmental Protection Agency, 226
enzyme, xiv, 7, 8, 10, 12, 15, 117, 119, 120, 121, 122, 128, 155
enzyme immobilization, 121, 155
enzymes, xiv, 9, 15, 16, 23, 29, 77, 94, 118, 119, 120, 121, 153, 154, 209
EPA, 212, 225, 226
epidermis, 158, 199
epithelial cells, 21, 23
Epoxidized soybean oil, xiii, 83, 84, 92
equilibrium, 3, 222
equipment, 39, 86, 87, 96, 140, 196, 197, 198
erosion, 8, 96
ester, 3, 5, 118, 120
ester bonds, 3, 5
ethanol, 39, 85, 88, 89
ethylene, 230
EU, 84, 163, 209, 215, 218, 221
eukaryotic, 24
Europe, 194
European Commission, 92
European Regional Development Fund, 204
European Union, 92, 216
evaporation, 27, 113, 125
evidence, 19, 74, 132, 154
evolution, xiii, 40, 64, 93, 96, 129, 130, 131
exclusion, 119, 126
experimental condition, 3, 123
exposure, 5, 76, 92
extracellular matrix, 22, 23
extraction, 85, 87, 135, 140, 210, 211, 215, 219, 221
extracts, 9, 15, 85, 86, 212, 213, 216, 217, 218, 223
extrusion, 39, 69

F

factories, 218
fat, 84, 85, 88, 89, 91, 118, 138, 140, 141
fatty acids, 52, 86, 118, 138, 145, 146, 147
FBI, 102
feedstock, 52, 68

fiber, 18, 54, 56, 57, 59, 219, 229, 232, 233, 236
fibers, 12, 18, 53, 54, 55, 56, 57, 59, 63, 123, 156, 210, 213, 219, 223, 228, 231, 232, 233, 236
fibroblasts, 17, 18, 19, 21, 22, 23
fibrosis, 20
fibrous cap, 15, 18
fibrous tissue, 19
filament, 2, 13, 20
fillers, 55, 60, 64, 170, 173, 178, 210
film formation, 188
film thickness, 4, 40, 69, 86, 112, 172, 183
films, xi, xii, xiii, xv, 2, 3, 4, 5, 6, 7, 8, 10, 11, 12, 13, 14, 15, 18, 21, 22, 23, 26, 27, 28, 38, 39, 40, 48, 67, 68, 69, 70, 72, 73, 74, 75, 76, 77, 79, 84, 93, 95, 96, 97, 105, 106, 107, 108, 109, 110, 112, 113, 114, 155, 156, 181, 182, 183, 184, 186, 187, 188, 189, 190, 191
filtration, 125, 209
financial, 79, 101, 114, 190, 204, 223, 236
financial support, 79, 101, 114, 190, 204, 223, 236
Finland, 70
fixation, 17, 18, 25
flax fiber, 228, 231, 232, 233
flexibility, 121, 188, 190
flocculation, 155
flora, 160
flour, 182
flow curves, 141
Flynn-Wall method, 47
food, xi, xiii, xv, 37, 39, 48, 68, 73, 74, 77, 78, 83, 84, 85, 87, 88, 89, 90, 92, 93, 106, 107, 114, 118, 138, 151, 153, 155, 156, 158, 170, 182, 183
food additives, 39
food chain, 77
food industry, 138, 155
food products, 106, 138, 156
footwear, xv, xvi, 152, 158, 162, 193, 194, 201, 202, 209, 219, 226
force, 84
Ford, 102

Index

formaldehyde, 215, 225
formation, 18, 19, 26, 43, 47, 69, 77, 95, 120, 131, 138, 154, 171, 190
formula, 122, 123
Fourier transform infrared (FTIR), xvi, 181
fragments, 15, 77, 78, 140, 154
France, 51, 53, 117, 237
free volume, 188, 190
fruits, 107, 143, 148
FTIR, xvi, 181, 184, 185, 186, 188
fungi, xiii, 9, 11, 75, 77, 94, 95, 96, 97, 99, 153, 158
fusion, 69

G

gasification, 208
gastrointestinal tract, 20
genes, 17
geometry, 3, 57
Germany, 53, 226
glass transition, 54, 77
glass transition temperature, 54, 77
glutamate, 120
glutamic acid, 185, 189
glycerol, xv, 70, 73, 76, 78, 181, 182, 183, 184, 185, 186, 187, 188, 189, 190
Glycerol-plasticized soya protein films, xv, 181
glycol, 221
grades, 228
grants, 29
granules, 213
Greece, 107
growing polymer chain, 120
growth, xiv, 11, 21, 22, 23, 95, 118, 131, 132, 135, 155, 156, 158, 160, 161, 162, 219, 226
growth rate, 11
guidance, 12, 17, 20

H

halogens, 209

hardness, 226
hazardous waste, 221, 223
hazardous wastes, 221, 223
HDPE, 74
HE, 32
healing, 19, 153
health, xi, 2, 138
heating rate, 47, 86, 141, 142, 146, 147, 171, 172, 229
heavy metals, 209
hemicellulose, 170
hemostasis, 24
hemp, 231
hepatocytes, 21
hexane, 85, 86
high fat, 138
high strength, 68
histidine, 120, 189
homogeneity, 159
human, xi, 21, 23, 37, 38
human health, xi, 37, 38
humidity, xiii, 68, 70, 71, 73, 77, 93, 94, 96, 100, 108, 134, 158, 211, 213
humus, 94
hyaline, 100
hybrid, xiv, 118, 124
hydrogels, 155
hydrogen, xii, 47, 67, 73, 84, 120, 124, 185, 188, 189, 190, 199
hydrogen bonds, 124, 188, 189, 190
hydrogen chloride, 84
hydrogen peroxide, 47
hydrogenation, 138
hydrolysis, 2, 3, 4, 6, 7, 8, 9, 11, 12, 14, 16, 28, 77, 118, 120, 209
hydroperoxides, 44, 45
hydrophilicity, xiii, 23, 68, 161
hydrophobicity, 10, 23, 78
hydroxide, 196
hydroxyapatite, 122
hydroxyl, xiv, 25, 73, 118, 120, 124, 130, 132, 134, 135, 182, 188, 190
hydroxyl groups, xiv, 73, 118, 124, 130, 132, 134, 135, 188, 190
hydroxytyrosol (HT), xi, 37

hygiene, 158
hypothesis, 154, 201

I

ideal, 25, 52, 59
identification, 86, 100, 175
image, 22, 173
images, xiv, 105, 109, 113, 162
immersion, 71, 76
immobilization, 121, 128, 154
immune response, 24
impact strength, xvii, 227, 228, 230, 231, 234
implants, 2, 14, 15, 18, 122, 154
improvements, xvii, 227
impurities, 14, 96, 119
in vitro, 2, 3, 7, 12, 14, 16, 21, 23, 28, 75, 157, 203
in vivo, 2, 12, 14, 15, 16, 28, 29
incubation time, 78
indentation, 226
indirect measure, 202
individuals, 158
induction, xiv, 38, 39, 42, 43, 44, 45, 48, 105, 108
induction time, 38, 39, 42, 43, 44, 45, 108
industrial processing, 208
industrial sectors, 39, 53, 155
industries, 64, 106, 158, 170
industry, xv, xvi, 84, 87, 91, 110, 152, 153, 158, 160, 161, 170, 193, 194, 195, 201, 202, 209, 228
infection, 158
inflammation, 18, 21, 28
inflammatory cells, 18, 20
ingredients, 106
inhibition, 29, 154, 156, 161, 162
inhibitor, 43
initiation, 12, 96, 130
injections, 15
innovative green polyamides (DAPA), xii, 51
inoculum, 159, 160, 161, 162, 196, 199
inorganic (nano)particles, xiv, 117, 122

insects, 152
insertion, 20
Instron, 55, 71, 109
insulation, 208
integration, 71
integrity, 73, 153
interface, 18, 19, 59, 77, 118, 120
intravenously, 12, 13
ion channels, 24
ionization, 86, 172
ions, 86
iron, xi, 101, 196, 199
isoprene, 210
isotherms, 62
issues, 23, 64

J

Japan, 40, 108, 109, 171

K

ketones, 43, 178
kidney, 9, 17, 24
kinetic equations, 28
kinetics, xi, 1, 3, 26, 28, 127, 134

L

laboratory tests, 11
lactic acid, xii, 67, 68, 70, 75, 81, 227, 236
lamination, 78
landfills, 215, 223, 226
leaching, xvi, 14, 23, 208, 211, 212, 215, 216, 221, 223
lead, xiv, 9, 19, 47, 76, 78, 101, 112, 119, 130, 137, 153, 215, 216, 221
legs, 13
leukemia, 28
life cycle, 213
light, 43, 72, 95, 96
light scattering, 72
lignin, xv, 152, 170, 171, 175, 176, 177, 178

lipases, xiv, 9, 14, 17, 23, 118, 119, 120, 121
lipid oxidation, 156
lipoproteins, 24
liquids, 16, 226
liver, 9, 15, 17
low temperatures, 42, 75, 110
lumen, 20
Luo, 164
lymphocytes, 18
lysine, 189

M

macromolecules, 24, 188
macrophages, 15, 16, 18, 19, 21, 22
magnesium, 101, 123, 196
magnitude, 141
majority, 120, 170
mammals, 24
management, 194
manganese, 101
manufacturing, 2, 11, 25, 27, 73, 210
marine diatom, 152
mass, xv, 3, 4, 5, 6, 7, 9, 13, 15, 45, 72, 77, 84, 85, 86, 95, 96, 109, 146, 169, 172, 208, 214, 216, 217
mass loss, 3, 4, 5, 6, 7, 9, 13, 15, 45, 146
mass spectrometry, xv, 169
material degradation, 175
material surface, 23, 113
materials, xii, xiii, xiv, xv, xvi, 18, 19, 23, 39, 41, 45, 46, 47, 48, 51, 52, 53, 55, 56, 57, 59, 64, 68, 69, 73, 76, 78, 80, 83, 84, 87, 89, 92, 95, 97, 105, 112, 114, 118, 119, 122, 124, 126, 158, 160, 169, 170, 171, 172, 174, 175, 178, 182, 188, 193, 195, 204, 205, 208, 209, 210, 211, 213, 215, 218, 219, 221, 223, 224, 226, 229, 235
matrix, xvi, 5, 10, 22, 27, 39, 53, 54, 56, 57, 59, 60, 106, 114, 119, 146, 208, 210, 216, 219, 223, 228, 236
matter, xii, 67, 71, 73, 99, 100, 216
measurement, xvi, 57, 76, 194, 195, 205

measurements, 70, 72, 109, 184, 215, 229, 236
meat, 107
mechanical properties, xiv, xv, 7, 11, 15, 16, 53, 56, 57, 64, 68, 69, 105, 114, 169, 171, 174, 175, 178, 181, 188, 189, 210, 228
mechanical testing, 188
media, 5, 17, 118, 119, 159
medical, 2, 12, 17, 18, 24, 25, 28
medication, 26
medicine, 25, 28
melt, 8, 26, 39, 54, 107
melt flow index, 107
melting, 41, 42, 54, 108, 109, 141, 142, 143, 144, 147
melting temperature, 54, 109, 142, 143, 144, 147
membranes, 17, 18, 20, 155
mesenchymal stem cells, 21
mesothelium, 20
metabolism, 12, 94
metabolites, 21
metals, xi, xiii, 94, 100, 101, 119, 215
meter, 226
methanol, 125, 132
methodology, 97, 100, 161
MFI, 107, 228
Mg^{2+}, 124
mice, 21, 28
microcrystalline, 228
microcrystalline cellulose, 228
micrometer, 70, 184
microorganism, 98, 157, 161
microorganisms, xiii, xvi, 11, 24, 75, 77, 79, 93, 94, 95, 98, 100, 157, 158, 160, 161, 162, 194, 195, 196, 202, 203
microscope, 97, 171, 229, 230
microscopy, 109, 154, 162
microspheres, 2, 5, 12, 13, 15, 20, 25, 26, 27, 28, 154, 155
microstructure, xii, 15, 23, 67, 230
migration, xiii, 83, 84, 85, 86, 87, 88, 91, 92, 107

Miguel Hernández University of Elche (UMH), xvi, 194
mitochondria, 24
mixing, 190, 213
modelling, 102
models, xii, 12, 20, 29, 52, 53, 58, 60, 64
modifications, 153, 196
modulus, xii, xiv, xv, 7, 10, 52, 53, 55, 56, 57, 58, 64, 68, 74, 78, 106, 114, 181, 188, 227
moisture, 68, 69, 71, 76, 77, 78, 94, 155, 183, 188
moisture content, 71, 94, 183, 188
molar volume, 214
mole, 214
molecular mobility, 188
molecular structure, 52
molecular weight, 2, 3, 4, 5, 7, 8, 10, 11, 12, 13, 14, 15, 17, 19, 24, 26, 27, 54, 70, 119, 120, 126, 128, 129, 130, 131, 132, 133, 134, 152, 157, 162, 163
molecular weight distribution, 5
molecules, xiii, 3, 68, 69, 76, 79, 94, 123, 124, 134, 188, 189
monolayer, 19
monomers, 8, 52, 96
morbidity, 158
morphology, xi, xiii, 1, 2, 5, 20, 23, 26, 68, 100, 113, 123, 124, 162, 170, 171, 172, 174, 229
Moscow, 1
moulding, 69, 228, 232, 233
mucosa, 23
mucous membrane, 154
mucous membranes, 154
multilayer films, 68, 69, 70, 73
multivariate analysis, 138
muscles, 9

N

Na^+, 123
nanocomposites, 122
nanometers, 71
nanoparticles, xiv, 118, 134, 154, 155

National Research Council, 79
natural resources, 68, 182
necrosis, 17, 18
nerve, 12, 13, 17, 20
Netherlands, 183
neutral, 17, 95
NH2, 189
nickel, 101
nitrates, 29
nitric oxide, 17, 25, 29
nitrile rubber, 210
nitrogen, xiii, 53, 71, 87, 94, 95, 96, 98, 108, 124, 125, 140, 199
NMR, xiv, 118, 125, 126, 127, 128, 129, 130, 131
non-polar, 154, 188
nuclear magnetic resonance, 125
nucleotides, 153
nutrient, 94, 95, 159
nutrients, 23, 158

O

octane, 85, 89
OH, 122, 123, 189, 202
oil, xi, xii, xiii, xv, 51, 63, 83, 84, 91, 92, 106, 118, 120, 125, 137, 138, 140, 141, 142, 143, 144, 145, 146, 147, 155, 182
oil samples, xv, 137, 138, 142, 143, 144, 145, 146, 147, 148
oligomers, 4, 8, 24, 75, 77, 157
olive oil, 38
Onset Oxidation Temperature (OOT), xii, 38
opacity, 71, 72, 77
operations, 20, 107
opportunities, 182
optical microscopy, 233
optical properties, 228
optimization, 40, 182
organ, 119, 123
organelles, 24
organic compounds, xvi, 100, 194
organic matter, xiii, 71, 93, 95, 96, 97, 99, 100, 216

organic polymers, 123
organic solvents, 120, 121
organism, 160
organs, 24, 25
osseus, 12
osteogenic sarcoma, 21
osteomyelitis, 29
oxidation, xiv, 39, 40, 42, 43, 44, 48, 95, 96, 108, 110, 137, 139, 145, 146
Oxidation Induction Time (OIT), xi, 38
oxidation rate, 95
oxo-biodegradable polymeric materials, xiii, 93
oxygen, xii, xiv, 40, 41, 42, 46, 47, 67, 68, 73, 78, 95, 105, 108, 112, 113, 114, 155, 199, 204
oxygen transmission rate (OTR), xiv, 73, 105

P

paclitaxel, 27
pain, 28
pancreas, 118
parallel, xiv, 18, 19, 56, 62, 108, 117, 121, 123, 159, 160, 228, 230
pathogens, 106, 156
PCT, 224
peptide, 73, 185, 189
peptides, 76
pericardium, 12, 13, 20, 21
periodontal, 17, 18
permeability, xii, 67, 73, 153, 162
permeation, 73, 108, 154
permit, 15, 20, 199
petroleum, xii, 51, 68, 140, 182
Petroleum, 107, 114
pH, xiii, 3, 4, 5, 6, 7, 8, 10, 17, 24, 71, 85, 93, 94, 95, 96, 99, 119, 157, 159, 163, 185, 212, 215, 216, 217, 225
phagocytosis, 17
pharmaceutical, xv, 25, 151, 153, 155
pharmacology, 25

PHB, xi, 1, 2, 3, 4, 5, 6, 7, 8, 9, 10, 11, 12, 13, 14, 16, 17, 18, 19, 20, 21, 22, 23, 24, 25, 26, 27, 28
phenol, 177
phenolic compounds, xi, 37, 177
Philadelphia, 80
phosphate, 3, 4, 5, 7, 10, 154, 196
phospholipids, 153
phosphorus, 199
phosphorylation, 94
photographs, 6, 8, 76
phthalates, 84, 87
physical and mechanical properties, 210
physical properties, 209, 218
physics, 59
physiology, 23
phytosterols, 138
plants, 38, 106, 211, 223
plasma membrane, 24
plastic deformation, 55
plastic products, 182
plasticization, xiv, 69, 87, 106
plasticized films, 70
plasticizer, xiii, xvi, 83, 84, 85, 90, 181, 188, 190
plastics, 68, 80, 92, 182, 195, 205, 230
platelets, 123, 132
PM, 34, 65, 164, 178
polar, 154, 182, 188
polarity, 182
pollution, 182, 194
poly(3-hydroxybutyrate), xi, 1, 2
poly(vinyl chloride), 84
polyamides, xii, 51, 52, 63
polycarbonates, 118, 120
polycyclic aromatic hydrocarbon, 178
polydispersity, 126
polyesters, xiv, 117, 118, 120, 122, 124
polymer, xi, xiii, xiv, 1, 2, 3, 4, 7, 8, 10, 11, 12, 14, 18, 19, 22, 23, 25, 26, 27, 37, 38, 39, 42, 43, 44, 45, 48, 64, 68, 71, 75, 78, 81, 84, 95, 96, 106, 107, 109, 110, 112, 113, 114, 117, 118, 119, 121, 122, 125, 127, 128, 132, 152, 154, 170, 173, 178, 214, 221, 226, 227

polymer chain, xiv, 3, 5, 8, 12, 78, 84, 118, 132, 152
polymer chains, xiv, 3, 8, 12, 78, 84, 118
polymer films, 3, 7, 23
polymer materials, 38
polymer matrix, xiii, xiv, 4, 10, 39, 45, 48, 64, 68, 106, 107, 113, 114
polymer molecule, 4
polymer oxidation, 45
polymer properties, 22
polymer structure, 112
polymer synthesis, xiv, 117
polymer systems, 25, 28
polymeric materials, xiii, 93, 106, 118
polymerization, xiv, 4, 52, 53, 117, 118, 119, 120, 121, 124, 125, 127, 128, 129, 130, 132, 134, 135
polymerization kinetics, xiv, 118, 128, 132
polymerization time, 128, 132
polymers, xi, xiii, xv, 25, 32, 34, 38, 42, 45, 47, 55, 59, 64, 68, 69, 93, 94, 119, 121, 123, 152, 169, 190
polymorphism, 90
polyolefins, 44, 45, 110, 228
polypeptides, 52
polyphenols, 138
polypropylene, xi, 38, 39, 43, 107, 121
polypropylene (PP), xi, 38, 39, 107
polysaccharide, 152
POOH, 44
pools, 160
population, 95
population growth, 95
porosity, xiv, 2, 105, 113, 174
Portugal, 151, 159, 207
potassium, 101
potential polymer stabilizer, xi, 38, 39
precipitation, 132
preparation, xiv, xv, 39, 69, 70, 100, 118, 124, 138, 147, 213
preservation, 138
preservative, 106
pressure gradient, 71
prevention, 20, 162
process control, 94

project, 163, 223
prokaryotes, 24
proliferation, 21, 22, 23
proline, 189
promoter, 155
propagation, 120
protection, 38, 138
protein structure, 188
protein synthesis, 154
proteins, 23, 24, 52, 69, 73, 152, 154, 182, 188, 189, 191, 209
protons, 128
prototype, 198, 199
prototypes, 12
pruning, 96, 170
Pseudomonas aeruginosa, 159
pulmonary arteries, 19
purification, 183, 209
purity, 5, 39
PVC, xiii, 83, 84, 87, 91
pyrolysis, xv, 169, 170, 171, 172, 175, 176, 177, 178, 208

Q

quantification, 86, 198, 202
quinone, 47

R

radiation, 155
radicals, 42, 43
rapeseed oil, xii, 51, 63
raw materials, xi
reactant, 183
reaction medium, 129, 130, 132
reaction rate, 128
reactions, 18, 44, 118, 119, 120, 121, 125, 127, 128, 129, 132, 138
reactivity, 18
reagents, 147, 212
recombination, 42
reconstruction, 122
recovery, xvi, 125, 208, 211, 217, 218

Index

recycling, xvi, 53, 207, 209, 210, 223
refractive index, 126
regeneration, 15, 19, 20, 21
reinforcement, 57, 210, 219, 233
relative size, 10
relevance, 162
relief, 28
repair, 17, 20
requirements, 95, 196, 198
researchers, 15, 203
residuals, 44, 170, 178
residues, 100, 119, 120, 182, 213, 216, 217, 218, 223
resins, 196
resistance, xvi, 11, 68, 69, 73, 112, 122, 190, 208, 212, 219, 223, 226
resolution, 70, 162, 184
resources, xii, 38, 51, 68, 79, 119, 170
response, 18, 19, 20, 21, 24, 55
restoration, 16, 19
restructuring, 190
retinol, 155
rheometry, 214
right atrium, 13
ring opening polymerization (ROP), xiv, 117
rings, 118
risk, 138
RNA, 21, 154
room temperature, 40, 53, 70, 72, 126, 159, 183, 216
roughness, 173
routes, 77
rubber, xvi, 125, 208, 210, 213, 214, 215, 218, 221, 223, 224, 226, 228
rubber compounds, xvi, 208
Russia, 1

S

safety, xi, 2, 114, 138
salts, 24, 123, 209
saturated fat, 145, 146
saturated fatty acids, 145, 146
sawdust, 171

scanning electron microscopy, 170, 172
sediments, 225, 226
seed, 138, 140
selectivity, 119
SEM analysis, xiii, 68
SEM micrographs, 235
semi-crystalline polymers, 231
sensitivity, 42, 160, 161
sepiolite clays, xiv, 118, 124
septum, 125
serine, 9, 17, 120, 189
serum, 9, 23, 138
sewage, 195, 204, 226
shape, 55, 229
shelf life, 106, 139
showing, 98, 106, 110, 127, 161, 162
side chain, 188
signals, 127, 128
signs, 15, 21, 28
silanol groups, 124
silica, 121, 122
simulation, 203
skeleton, 185
skin, 17, 154, 156, 158, 160, 199
Slovakia, 40
sludge, 11, 195, 204, 213, 218, 224, 226
smooth muscle, 19, 21
smooth muscle cells, 19, 21
smoothing, 23
society, 152
sodium, 69, 101, 124, 209, 210, 212
software, 173, 229, 230
solid waste, 96, 194
solubility, 27, 152, 153, 157
solution, xvi, 5, 6, 7, 9, 12, 85, 95, 120, 124, 125, 132, 155, 159, 160, 161, 183, 188, 193, 195, 208, 210, 211, 213, 215, 217, 223
Soxhlet extractor, 233
Spain, 30, 37, 39, 83, 86, 90, 105, 107, 108, 124, 137, 139, 140, 169, 171, 181, 193
species, 11, 43, 94, 120
specific migration limits (SME), xiii, 83
specific surface, 122, 123
spectroscopy, xvi, 181

spinal cord, 20
spinal cord injury, 20
spindle, 20
stability, xv, xvii, 38, 42, 45, 89, 110, 112, 137, 138, 144, 227, 228, 235
stabilization, xiv, 38, 48, 84, 89, 94, 105, 110, 112, 114
stabilization efficiency, 112
stabilizers, xi, 37, 38
standard deviation, 101
starch, 69, 76, 152
state, 8, 42, 71, 123
statistics, 140, 144
steel, xi, 54, 70
stereomicroscope, 171, 172
stereospecificity, 24
stomach, 20
storage, 73, 107, 158
stress, xii, 52, 53, 55, 59, 60, 62, 63, 64, 109, 157, 229
stress-strain curves, 55, 59, 109
stretching, 185
strong interaction, 59
structural defects, 44
structure, xi, 1, 4, 11, 20, 52, 53, 78, 120, 142, 146, 147, 153, 162, 171, 172, 173, 174, 175, 178, 190, 194, 209, 219, 221
substitutions, 123
substrate, 11, 120, 158, 201, 202
substrates, xiii, 23, 93, 94, 121
sulfate, xvi, 20, 207, 209, 223
sulfuric acid, xvi, 208, 209, 212
sulphur, 199
Sun, 191
supplementation, 156
surface area, 123
surface chemistry, 23
surface energy, 23, 123
surface layer, 8
surface properties, 23
surface structure, 2
survival, 21
susceptibility, 69, 75
suspensions, 157
sustainable development, xii, 51, 53, 64

suture, 13, 18
sweat, 158
Sweden, 237
swelling, 123, 215, 222
Switzerland, 86, 108, 140
synergistic effect, 112
synthesis, xiv, 24, 53, 85, 117, 118, 119, 120, 122, 135
synthetic polymers, 96

T

tacticity, 8
TCC, 159
TDI, xiii, 83
techniques, 23, 39, 42, 69, 108, 114, 138, 139, 147, 170, 171, 195
technologies, 2, 12, 139, 194
technology, 81
TEM, 16
temperature, xii, xiii, xvi, 3, 4, 40, 45, 47, 51, 54, 59, 60, 61, 62, 70, 71, 77, 86, 94, 96, 98, 99, 108, 109, 119, 134, 140, 141, 143, 146, 147, 157, 171, 172, 196, 198, 208, 211, 215, 217, 228, 229
tensile strength, xv, xvi, 7, 10, 69, 74, 78, 181, 187, 188, 189, 190, 208, 219, 223, 228
tension, 219
testing, xiv, 40, 42, 55, 71, 80, 97, 98, 99, 100, 101, 105, 163, 183, 184, 196, 198, 201, 203, 229, 230
textiles, 155, 194
TGA, xii, xiv, xv, 38, 41, 45, 46, 86, 89, 105, 108, 110, 118, 125, 126, 132, 133, 137, 138, 139, 140, 142, 144, 145, 146, 147
therapy, 25, 154
thermal analysis, 47, 138
thermal degradation, xv, 110, 139, 147, 170, 171
thermal properties, 229, 233
thermal stability, xiv, 47, 105, 110, 121, 137, 138, 228
thermal treatment, 138

thermograms, 140
thermogravimetric analysis, xii, 38, 124, 126, 132
thermoplastics, 19, 53, 69
thin films, 13, 14
threonine, 189
thrombomodulin, 21
time use, 182
tissue, 9, 14, 15, 16, 18, 19, 20, 21, 23, 24, 28, 29, 71, 72, 154, 156
TNF, 17
TNF-α, 17
tocopherols, 38, 138
tolerable daily intake (TDI), xiii, 83
toluene, 124, 125, 214, 215, 222
toxic effect, 17
toxicity, xiii, xiv, 20, 28, 83, 84, 117, 152, 153
trade, 153
transesterification, 118, 128, 132
transformation, 94
transformations, 171
transition metal, 95
transition period, 84
transmission, xiv, 72, 73, 105, 108, 112, 113
transparency, xii, 67, 69, 72, 228
transport, 94
transportation, 73
treatment, 5, 17, 18, 23, 28, 41, 147, 156, 159, 160, 195, 199, 209, 210, 213, 215, 218, 219, 223
tricarboxylic acid, 94
tricarboxylic acid cycle, 94
triglycerides, 118, 120, 145
tuberculosis, 29
tumor, 17, 21
tumor necrosis factor, 21
Turkey, 74, 117
tyrosine, 189

U

UK, 91, 109
underlying mechanisms, 153
uniform, xii, 16, 19, 25, 26, 28, 67

United, 226
United States, 226
urban, 96, 100, 195
urea, 210
USA, xiv, 40, 54, 86, 87, 107, 108, 124, 137, 140, 171, 172, 191
UV, 71, 126

V

vacuum, 72, 125, 211
Valencia, 37, 169, 171
vapor, 73
variables, xvi, 147, 181, 203
variations, 45, 157, 162
vascular diseases, 29
vasodilator, 25
vegetable oil, xiii, 38, 52, 83, 84
vegetables, 107, 156
vessels, 24
vibration, 185
vinyl chloride, xiii, 83
vinyl monomers, 209
viscosity, 12, 152, 157, 213, 221
volatilization, 110, 171
vulcanizates, 210, 215, 221, 222
vulcanization, 210, 214, 221

W

waste, xvi, 95, 170, 193, 194, 195, 196, 208, 209, 210, 212, 213, 215, 216, 219, 221, 225, 226
waste management, 194
waste water, 226
wastewater, 196, 199
water, xii, xvi, 3, 4, 8, 11, 24, 26, 27, 68, 70, 71, 72, 73, 76, 77, 78, 79, 85, 94, 95, 118, 119, 120, 122, 124, 130, 134, 146, 155, 183, 190, 195, 208, 211, 218, 221, 223, 226
water absorption, xii, 68, 71, 76, 77
water sorption, 72
weight gain, 76

weight loss, xii, 5, 14, 15, 26, 28, 68, 72, 76, 110, 132, 134, 144, 146, 147, 196, 203
weight ratio, 27
wood, xv, 169, 170, 171, 174, 175, 177, 178
wood species, 171
wool, 155
workers, 154, 157, 161
worldwide, 170
wound healing, 156

X

XRD, 78

Y

yeast, 75, 100
yield, xii, xiv, 52, 53, 55, 59, 60, 62, 63, 64, 106

Z

zinc, 101
ZnO, 210